HYDROLOGICAL DIMENSIONING AND OPERATION OF RESERVOIRS

HYDROLOGICAL DIMENSIONING AND OPERATION OF RESERVOIRS

Practical Design Concepts and Principles

by

IMRE V. NAGY

Budapest Technical University & Committee for Water Resources Development, Hungarian Academy of Sciences, Hungary

KOFI ASANTE-DUAH

Anteon Corporation, Environment Division, San Diego, California, U.S.A.

and

ISTVAN ZSUFFA

Department for Hydrology and Water Management, Budapest Technical University, Hungary

Springer-Science+Business Media, B.V.

A C.I.P. Catalogue record for this book is available from the Library of Congress.

ISBN 978-90-481-5942-0 ISBN 978-94-015-9894-1 (eBook)
DOI 10.1007/978-94-015-9894-1

Cover illustration: Network flow representation of spatial and temporal configuration for multireservoir systems

Printed on acid-free paper

To Our Extraordinary Families

TABLE OF CONTENTS

Preface xiii

CHAPTER 1: INTRODUCTION 1

1.1. Water Resources Management 2
1.1.1. Planning and Design of Water Resources Systems 3
1.1.2. Water Resources Systems Analysis 3
1.1.3. Why Build Dams? 4
1.2. The Need for Flow Regulation Systems 5
1.2.1. Balancing Water Supply with Water Demands 5
1.2.2. The Water Balance of Reservoirs as a Tool in Reservoir Design and Management 8
1.2.3. Regional Imbalance Between Water Supply and Water Demand: Transboundary River Management Issues 10
1.3. Impact of Reservoir Projects on the Temporal and Spatial Variation of Streamflow Quantity and Quality 11
1.3.1. Environmental Impact Issues Associated with Dam and Reservoir Projects 11
1.4. Using Storage Reservoirs in Flow Regulation and Water Management Schemes 13

CHAPTER 2: THE HYDROLOGY OF FLOW REGULATION 15

2.1. Flow Allocations 15
2.2. Hydrological Problems Resulting from Water Deficiency 16
2.3. Hydrological Problems Caused by Water Excess 17
2.4. Types of Flow Regulation 19
2.4.1. Daily Flow Regulation 19
2.4.2. Weekly Flow Regulation 20
2.4.3. Seasonal or Annual Flow Regulation 22
2.4.4. Multiannual Flow Regulation 22
2.5. Optimizing Flow Regulation Schemes 25
2.6. General Characteristics of Flow Regulation Systems 25

CHAPTER 3: PLANNING FOR DAMS AND RESERVOIRS: HYDROLOGIC DESIGN ELEMENTS AND OPERATIONAL CHARACTERISTICS OF STORAGE RESERVOIRS **29**

3.1. The Reservoir Design Problem **30**
3.1.1. Reservoir-Site Selection 31
3.1.2. The Case for Multipurpose Reservoirs 32
3.2. Multireservoir System Layout and Analyses **33**
3.3. Hydrological Basis for the Determination of Reservoir Storage Capacity **35**
3.3.1. A Model of Capacity Allocation and Survey of Water Demands in Multipurpose Reservoirs 37
3.3.2. Estimating the Active Storage Necessary for Flow Regulation and Water Supply 40
3.3.3. Hydroelectric Power Potential of Storage Reservoirs 41
3.3.4. Storage-Space for Flood Mitigation: The Reservoir Flood Storage Capacity Design 41
3.3.5. Siltation of Reservoirs and Sediment Reserve Storage 44
3.3.6. Adjustment of Storage Estimates for Net Evaporation Losses 48
3.3.7. Other Secondary Factors Affecting Reservoir Size-Selection 51
3.4. Hydrologic Data Requirements and Analyses **53**
3.4.1. Selecting a Distribution for Use in the Hydrologic Design Process 54
3.4.2. Bayesian Techniques for Parameter Estimation with Limited Data 55
3.4.3. Design of Reservoir Storage for Stochastically Varying Water Demand 57
3.5. Deterministic vs. Stochastic Methods in the Reservoir Design Problem **57**
3.5.1. Deterministic Methods and Models in Reservoir Design 58
3.5.2. Stochastic Problems in the Design of Reservoirs 58
3.6. Guidelines for the Hydrological Dimensioning of Reservoirs **59**

CHAPTER 4: PRINCIPLES AND CONCEPTS IN THE HYDROLOGIC DESIGN AND OPERATION OF STORAGE RESERVOIRS **61**

4.1. Utilization of Reliability-Based Techniques in the Hydrologic Design Process **61**
4.1.1. The Concept of Reservoir Efficiency Functions 62
4.1.2. The Efficiency Function as a Basis for Storage Determination 65
4.1.3. General Types of Reliability Parameters vs. Reservoir Efficiency Functions 68
4.1.4. Storage Allocation in Multipurpose Reservoirs 71
4.2. Topographical Characteristic of the Reservoir **72**
4.3. Modeling Methodology for the Systems Simulation of Reservoir Design Problems **75**
4.3.1. A Flood Storage Submodel 80
4.4. Optimal Release Policies in the Operation of Multipurpose Reservoirs **81**
4.4.1. Stochastic Approach to Establishing an Optimal Release Policy 82
4.5. The Value of Hydrologic Information in the Management of Reservoirs **83**

CHAPTER 5: SYSTEMS APPROACH IN THE HYDROLOGIC DESIGN AND OPERATION OF STORAGE RESERVOIRS **85**

5.1. Hydrologic Models in Water Resource Systems **85**
5.1.1. Time Series Models in Hydrologic Modeling 88
5.1.2. Modeling Techniques 89
5.2. Design of Reservoir Storage-Capacity with Inadequate Hydrologic Data **90**
5.2.1. The Need for Synthetic Data: Stochastic Generation of Synthetic Data and Flow Generation Strategies 91
5.2.2 Multivariate Stochastic Models 92
5.3. Evolution and Philosophy of Stochastic Simulation in Reservoir Systems Modeling and Design **93**
5.4. Deterministic-Stochastic Hybrid Models **95**
5.5. Risk and Uncertainty in Reservoir Design **95**
5.5.1. Model Verification and Performance 96
5.5.2. Reservoir Performance Reliability 97
5.5.3. Uncertainty Assessment via Sensitivity Analysis 99
5.6. The Hydro-Economics of Reservoir Design **100**
5.6.1. Economic Concepts in Reservoir Planning 100
5.6.2. Benefit-Cost and Cost-Effectiveness Analyses *versus* Optimal Design from Marginal Analysis 101
5.7. Optimization Techniques as a Design Tool for Water Resource Systems **102**
5.7.1. The Optimization Problem 102
5.7.2. Solution Techniques for the Optimization Problem 103
5.7.3. Multiobjective Optimization in Reservoir Design 104
5.7.4. The Net-Benefit Function Under Optimality Conditions 105
5.8. Optimization Under Uncertainty and Risk **106**

CHAPTER 6: HYDROLOGIC ANALYSIS OF FLOOD FLOWS **109**

6.1. Design Flood Determination **109**
6.1.1. The Selection of an Acceptable Risk Level 110
6.1.2. The Calculation of the Risk of Overtopping 111
6.1.3. Hydro-Economic Impact Analysis 112
6.1.4. Modeling the Flood Flows 113
6.2. The Probability of Occurrence of Flood Flows **115**
6.2.1. Estimation of the Probability Distribution Function of the Maximum Flood Flows for Large Rivers 116
6.2.2. Estimation of the Probability Distribution Function of the Maximum Flood Flows for Small Rivers 118
6.2.3. Estimation of the Probability Distribution Function of Maximum Floods in the Case of Medium-sized Rivers 120
6.3. Estimation of Flood Flows Using Limited Data **122**
6.4. Estimation of Flood Attenuation by Reservoirs **122**

CHAPTER 7: METHODS OF APPROACH FOR DESIGNING OPTIMAL STORAGE CAPACITIES AND OPERATIONAL STRATEGIES FOR MULTIRESERVOIR SYSTEMS 125

7.1. **A Review and Classification of Reservoir Capacity-Yield Estimation Procedures** **125**
7.2. **Critical Period Techniques** **126**
7.2.1. Reservoir Capacity-Yield Estimation by Mass-curve Procedure 127
7.3. **Probability Matrix Methods** **129**
7.3.1. Basic Principles of Reservoir Sizing Using Probabilistic Methods 130
7.4. **Reservoir Storage Requirements from Stochastic Data** **131**
7.4.1. Basic Elements of Stocahstically-Generated Data 132
7.5. **Choosing Between Deterministic vs. Probabilistic vs. Simulation Methods** **137**

CHAPTER 8: DETERMINATION OF THE OPTIMAL RESERVOIR STORAGE CAPACITY AND OPERATIONAL PARAMETERS FOR A RIVER DAM 139

8.1. **Screening Models for Multireservoir Systems Design** **140**
8.1.1. A Cost-Efficient Reservoir Capacity Design in Multireservoir Systems 141
8.2. **Reservoir Network Analyses for Model Development** **142**
8.3. **Formulation of the Reservoir Model and Modeling Methodology** **146**
8.3.1. Using Monte Carlo Techniques in the Optimal Design of Reservoir Systems 151
8.4.1. **A Multisite Multiseason Flow Generation Strategy** **151**
8.4.1. The Multivariate Autoregressive AR(1) Model for the Multisite Annual Generation Scheme 152
8.4.2. The Disaggregation of Annual Streamflow Data 156
8.4.3. Implementation of the Flow Generation Algorithm 157
8.5. **Storage-Capacity Allocation to Reservoir Sites** **157**
8.5.1. Determination of the Reservoir Sizing Factors 158
8.5.2. Incorporating a Reliability Measure 159
8.5.3. Matrix of 'Disaggregated' Water Demands 160
8.5.4. Matrix of Storage Volumes 162
8.6. **Cost-Efficient Capacity Allocations in the Design of Multireservoir Systems** **163**
8.7. **An Optimal Solution for the Multireservoir System Design** **165**
8.7.1. The Overall Model Implementation Process 166
8.8. **Optimal Storage Capacity Decisions for Multireservoir Systems** **167**

CHAPTER 9: HYDROLOGICAL SIZING OF RESERVOIRS FOR FLOOD PROTECTION 169

9.1. **Determination of the Characteristic Hydrograph** **169**

9.2. **Spillway and Sluice Gate Considerations in the Design and Operation of Flood Retention Reservoirs** **170**
9.2.1. The Case for a Regulated Sluice 171
9.2.2. The Case for a Closed Sluice 174
9.2.3. The Case for an Opened Sluice 174
9.2.4. Comparison of the Dimensioning Methods 177
9.3. **Approximate Evaluation of the Efficiency Function** **179**
9.4. **Emergency Flood Storage** **182**
9.4.1. Impacts of Emergency Storage in the River System 183
9.5. **Design and Operation of an Emergency Flood Control Program** **184**

CHAPTER 10: APPLICATION OF THE MORAN MODEL IN RESERVOIR STORAGE DESIGN **187**

10.1. **The Moran Model** **187**
10.2. **A Proposed Mathematical Model** **190**
10.2.1. Determination of the Transition Probabilities 192
10.2.2. Model Application - An Example 201
10.3. **A Probabilistic Model for the Determination of the Reservoir Efficiency Function** **204**
10.3.1. The Basic Hypotheses 204
10.3.2. The Basic Relationships 205
10.3.3. Determination of the Transition Probability Matrix 206
10.3.4. Determination of the 'Behavior Function' 207
10.4. **Concluding Remarks** **208**

LIST OF REFERENCES AND BIBLIOGRAPHY **209**
Literature Cited 209
Additional Suggested Literature 217

Index **221**

PREFACE

Storage reservoirs represent one of the most effective tools for eliminating, or at least for minimizing, discrepancies in the time and space variations of water resources distribution and requirements. In fact, the different – often contradictory – and increasing demands on water resources utilization and control usually can be fulfilled only by building multi-purpose reservoir systems. In this way, the available water resources can be exploited and/or managed in a more rational way. Typically, the construction of a dam across a river valley causes water to accumulate in a reservoir behind the dam; the volume of water accumulated in the reservoir will depend, in part, on the dimensions of the dam. The size of the dam will normally affect the capital expenditure in a very significant way. Indeed the construction of large water resource control systems – such as dams – generally involves rather huge manpower and material outlays. Consequently, the elaboration of effectual methods of approach that can be used in establishing the optimal reservoir parameters is of great practical significance. For instance, in the design and operation of large multi-reservoir systems, simple simulation and/or optimization models that can identify potentially cost-effective and efficient system design are highly desirable. But it should be recognized that the problem of finding optimal capacities for multi-reservoir systems often becomes computationally complex because of the large number of feasible configurations that usually need to be analyzed.

This book elaborates a systematic approach to the hydrological design and operation of reservoir storages that can be employed to help optimize the reservoir design process. It calls for the use of contemporary techniques and tools – especially for project planning; selection of reservoir site locations; water resources allocation and systems operation; best management practice decision-making; water quantity and quality assessments; hydrological monitoring studies; flood forecasting and flood control assessments; etc. Indeed, the application of novel techniques to hydrological problems are always very important – especially because problems often encountered in hydrological designs tend to require repeated sampling of the variables of interest, and that could become rather expensive. The novel methods of approach allows for the acquisition of information over large areas, and also for the better integration of all such data in a more cost-efficient manner. In addition, these methods of approach can allow a water resources system designer or analyst to engage in a variety of scenario evaluations that satisfy different project conditions – which can therefore facilitate an optimization of the overall system design and operation. By using the systematic approach offered by this title in a reservoir system design effort, different policies can be evaluated and compared, which will then facilitate the selection of an optimal design and operation plan for a proposed or existing dam project.

Overall, this book attempts to provide a concise, yet comprehensive overview of the hydrological dimensioning aspects relating to the design and operation of river dams. It elaborates some very important concepts, tools and methodologies that can be used to help resolve multiple reservoir capacity allocation problems in a consistent, efficient and cost-effective way. The material presented in the book will guide the hydrological analyst or design engineer in the determination of the optimal storage capacities and operational strategies of the individual elements in a system of reservoirs present at a set of potential dam sites. The specific goal in the type of optimization problem elaborated here will be to help minimize the reservoir capacities and/or shortfalls (which are a surrogate for minimizing costs) and still meet capacity requirements and needs. The ultimate objective in such an application will be to prevent over-design or under-design of reservoir storage capacities under optimum conditions of reservoir operation. Thus, the optimal solution will produce the smallest reservoir storages required to meet all project demands with an acceptable degree of reliability. The subject matter of this book should be of interest to practising hydrologists, water resources systems analysts and civil engineers, as well as to students taking a variety of courses in water resources planning and development.

We are indebted to a number of people for both the direct and indirect support afforded us during the period that we worked on this book project. Sincere thanks are due to our families, and several friends and colleagues who provided much-needed moral and enthusiastic support throughout preparation of the manuscript for this book. We thank the Publishing, Editorial and Production staff at Kluwer Academic Publishers (Dordrecht, The Netherlands) who helped bring this book project to a successful conclusion – with special thanks to Petra D. van Steenbergen (Publishing Editor); thanks also to several colleagues at Anteon Corporation, especially Max Voigtritter (Senior Program Manager) for providing miscellaneous support. We also wish to thank every author whose work is cited in this volume, for having provided some pioneering work to build on. Finally, it should be acknowledged that this book also benefited greatly from review comments of several anonymous individuals, as well as from discussions with a number of professional colleagues. Any shortcomings that remain are, however, the sole responsibility of the authors.

Imre V. Nagy, Budapest, Hungary (August, 2001)
Kofi Asante-Duah, San Diego, California, USA (August, 2001)
Istvan Zsuffa, Budapest, Hungary (August, 2001)

Chapter 1

INTRODUCTION

Why the need for reservoirs in water resources management programs? Simply, for the regulation and control of natural water flow patterns and regimes – in order to meet and satisfy the spatial and temporal water needs of society. In fact, prospects for improved standards of living in many regions of the world hinges in part on the development and rational management of water resources systems – including the construction of dams and/or reservoirs. Flow regulation in water resources management is indeed one of the most important tools for supporting irrigation to boost food production; for the protection of life and property by mitigating the destructive effects of floods and droughts; and for the generation of energy to supply electricity – among other things. Some of these benefits cannot be realized without the construction of dams and, subsequently, the effective operation of the concomitant storage reservoirs.

In this day and age, the increasing demand for sufficient quantities and quality of water has forced planners to contemplate complex design and operational strategies for water resource systems – that typically include the regulation of natural flow; the control of floods, droughts, and water quality conditions; and also the effective management of the risk, economic, political and social factors associated with such projects. Specific to storage reservoirs, the role of hydrology in reservoir regulation comes into play during the early design of a project and continues throughout its actual operation. Principally, the design of a reservoir requires knowledge of the quantity of streamflow and its occurrence with respect to area and time, and the operation of a reservoir requires the analysis of streamflow based on pre-reservoir flow records and on current streamflow and precipitation estimates. Ultimately, a successful method of approach best suited for the hydrological dimensioning and operation of storage reservoirs requires that the planners, designers, and other systems analysts grasp the issues and concerns that are important to all stakeholders.

This title offers key fundamental principles and concepts that will help in the efficient design of water control systems – in particular, making it possible to develop appropriate reservoir design and operational management decisions. The book introduces readers to some effectual methods of approach to utilize in the hydrological dimensioning and operation of storage reservoirs. The methods encompass both deterministic and probabilistic procedures. It is noteworthy that the most useful deterministic methods of analysis are often the simplest because they are easier to understand – but this does not suggest that the more sophisticated probabilistic methods are less useful. Indeed, an appropriate mathematical system of analysis typically is a compromise between the required level of details and a resulting potential for increased accuracy versus expected savings in model development time or cost and computational requirements. To instill some sense of balance and pragmatism between the competing desirable goals, we have attempted to organize the discussions on the various methods for evaluating and comparing alternative solutions in a form useful for teaching and self-study; the methods of approach that we present here have also seen successful field applications.

1.1. Water Resources Management

The prime objective of most public water resources development projects is often stated as the maximization of national welfare. To achieve this goal, contemporary water resources projects are mostly planned for more than one purpose, especially because such multi-purpose projects usually have greatly improved economic justification. The design, and subsequently the operation, of multipurpose reservoir systems is a complex undertaking that involves consideration of a large number of physical, economic, social, as well as hydrological and other engineering constraints. The problem is further complicated by the fact that, usually, the reservoirs are not independent – since they often act as components in an integrated system to regulate flows in downstream reaches of a water body.

A very important aspect of water resources management that could easily be overlooked during the early conceptualization stages of the project relates to a careful identification of all stakeholders. For example, in the past, there was almost no opposition to the planning and construction of navigation and hydropower schemes in much of Europe (viz., Switzerland, Germany, France, Austria, etc.). However, nowadays, no project can be realized without great delays resulting from innumerable objections, requests, and suggestions for changes and modifications – especially in relation to potential environmental effects. An example is the case of the joint Hungarian-Slovak Gabcikovo-Nagymaros project for which the primary goal was the improvement of navigation through the already existing Rhine-Maine-Danube waterway going from the Atlantic to the Black Sea. Although the importance of this transcontinental waterway to boost East-West trade cannot be overemphasized, it is unfortunate that the ongoing construction had to be stopped following opposition from some environmentalists. Indeed, this type of problem underscores the importance of organizing multidisciplinary (sometimes international) forums where those developing and planning such a project will meet with the stakeholders or interest groups – to allow for stakeholder and public comment and participation in the very early stages of project planning.

Ultimately, the potential benefits of river basin management programs typically include improvements such as river regulation, water supply, hydropower generation, flood control, recreation and navigation, fisheries, and water quality control. This complex management scheme is manipulated, invariably, by reservoir constructions, which have to be planned and operated in accordance with some prescribed optimization rules and objectives. However, due to the complexity of such river basin reservoir systems, it is not always easy to satisfy all of the possible design and operating policies; quite often, there is no assurance that the best combination of policies and structural measures can be found, especially when the river basin is extremely large with a large number of reservoirs or reservoir systems, and furthermore when a great variety of short- and long-term interests need to be met. Under such circumstances, a classical path to follow is to find a compromise control policy for the different interests.

1.1.1. PLANNING AND DESIGN OF WATER RESOURCES SYSTEMS

Water resources management is impossible without planning. The planning process is often described as an identification of the goals and the development of the information and data necessary to evaluate alternatives and recommend a course of action while providing information about the consequences of alternative plans that aid the decision-makers (Meta Systems, Inc., 1975). Kuiper (1971) defines planning as an organized attempt to select the best alternative to achieve a specific goal; this definition has a number of important implications. The general types of actions that take place in the water resources development process – from its earliest inception of an idea, through the stages of planning, design, and construction to the final management of a project is presented by Kuiper (1971).

The aim of any system design is to select the combination of variables that would tend to maximize the net benefits, within the constraints of the requirements of the design criteria. Under proper management the structural and nonstructural elements of a water-resource system mobilize the latent usefulness of natural water resources while allaying their potential destructiveness. The systems manager can only fulfill her/his task efficiently when s/he has a good planning system at her/his disposal; s/he can then approach optimal water resources management and make effective decisions. From the social point of view, an optimal situation might be reached when the different national, regional and local interests are weighed in a democratic way; this requires a clear presentation of the plan, its alternatives, and the social consequences.

It is noteworthy that, following the construction of storage reservoirs, there usually will be a change in the conditions of surface and subsurface runoff, environmental characteristics, and the earlier economic activities that existed at the project locale. Consequently there is a need to incorporate a detailed environmental impact assessment in the project analysis.

1.1.2. WATER RESOURCES SYSTEMS ANALYSIS

A system is a combination of units or component objects that interact in a regular manner to accomplish a set of objectives. Components of a water resource system may have physical, political, economic, legal, sociological, and/or environmental dimensions. Using mathematical and computer techniques, the system can be optimized, and the effect of changes in system parameters on the optimal solution can be investigated; this latter study – referred to as sensitivity analysis – often is a very important element in the ultimate decision-making step.

Increasingly, complicated decisions arising from the diversified interests in reservoir use have to be made by the system manager; knowledge of the trade-offs resulting from such decision will be helpful in the decision-making process. In general, multi-objective situations introduce complications; nevertheless, it has distinct advantages over most traditional planning approaches. Fiering *et al.* (1971) elaborates on the optimization between water resources assessment and requirement, drawing a distinction between when a requirement is considered as an objective and then as a constraint.

It should be noted here that, in general, matching the water management requirements with the topographic and hydrologic characteristics of a watercourse is an optimization problem that usually involves a search for the economic optimum. This optimum will typically be a function of various natural conditions as well as technical, economic, social, and environmental factors. Also, whereas the natural average water supply conditions may be approximately stationary in a statistical sense, the social, economic and environmental factors, as well as the cost of construction, operation and maintenance, and key investment decisions may undergo an order of magnitude of change with time. Consequently, the so-determined optimum may be valid only for a specified period of time.

In recent years water management has become more complicated due to increased water quantity and quality demands; more diversified and often conflicting interests in water use; and a growing public concern about environmental quality. For example, it is recognized that the impoundment of water in large and deep reservoirs is producing, in some instances, deleterious effects on water quality – albeit the interrelationships between reservoir system operations and system water quality regimes are very complex. Some of the more severe impacts of flow control on water quality occur during periods of low or zero releases from reservoirs; future streamflow requirements for water quality maintenance in the affected river reaches may require changes in the release schedules of controlling dams and thereby produce considerable changes of the present water quantity and quality regime in parts of the system. As a consequence, more and sometimes severely limiting constraints have been imposed by which the water resources system must be operated.

1.1.3. WHY BUILD DAMS?

Water is one of the most important natural resources available to humankind. But there has to be the right amount of water in the right place at the right time – often realized by building dams to regulate the natural flow regimes of available waters. The purpose of a dam is to impound/store water for a variety of reasons. Indeed, dams provide several benefits to society – including flood control, drinking water supply, irrigation water for farming, recreational areas, and clean renewable energy through hydropower; many dams fulfill a combination of these functions.

For storage reservoir projects, large-volume reservoirs (viz., having volume more than 10 mill. m^3) are usually designed for multipurpose use, and the smaller reservoirs are designed to satisfy only specific single-purpose needs (e.g., irrigation, energy production, industrial and drinking water supply or flood retention). In situations involving interconnected, jointly operated reservoirs, it is possible to have the upper reservoirs being used only to regulate the flow and the lower reservoirs used for the water supply. It is also noteworthy that, sometimes, some smaller reservoirs are constructed between the largest ones in order to improve the undisturbed regime and use of the lower reservoir.

Human-made dams may be classified on the basis of the type of construction material used, the methods used in construction, the slope or cross-section of the dam, the manner in which the dam resists the forces of the water pressure behind it, the means used for controlling seepage, and sometimes, according to the purpose of the dam (Jansen, 1988; Linsley and Franzini, 1979). The selection of the best type of dam for a given site and/or situation is very much dependent on the engineering feasibility and costs.

1.2. The Need for Flow Regulation Systems

In the past, simply adapting to natural flow regimes was the norm for societies. Thus, in consideration for human settlements, locations that were believed to be free from floods but where drinking water would be easily available were purposefully selected for developing human habitats. It was in the course of the demographic explosion during the 18th and 19th Centuries that the present forms of classic water management systems started taking shape via the construction of canals and/or the diking of rivers. Channels of watercourses were also enlarged to assure a better conveyance of floods, and the larger demands for water were met by means of water transfer to greater distances. The overall picture under this set of scenario is illustrated in Figure 1.1.

Increasing world populations, together with the quest for improved living standards and the protection of life and property created new water resources management problems, especially with respect to our attempts to match water supply with water demand. Alongside the continuing demographic explosion of the 20th Century came the development of water management systems that involved storage reservoirs (Figure 1.2). Instead of adaptation of water supply and demands to random processes, the transformation of water regime by storage has proven to be a reasonable and rational solution. In fact, this seems to be the only rational way to achieve the proper development and management of the varying nature of water resources found on the different continents and islands, or for the individual regions and countries of the world.

Although water is a renewable resource, and surface watercourses provide some 95 percent of fresh water, available waters tend to fluctuate in accordance with the random processes of the weather. Consequently, and ideally, water users must adjust their utilization patterns to be in line with these stochastic processes that typically is comprised of random sequences of high flow and low flow periods – that is if flow regulation is to be avoided.

1.2.1. BALANCING WATER SUPPLY WITH WATER DEMANDS

Water management using storage consists of an adaptation of random processes of water supply to human needs. Because of its random character, shortfalls in water supply that occur during dry seasons is usually compensated for by waters stored from wet seasons. Thus, the storage reservoirs built for flood control usually are emptied immediately after the flood has passed in order to enable them to receive the next flood. If there is a reliable forecast for the last flood, then under the known set of circumstances, the reservoirs may not be emptied in order that this can guarantee anticipated water withdrawals for the dry season. This process certainly constitutes a special optimization problem, whereby the decision maker has to take into account the potential benefits and losses based on the reliability of the forecast. Indeed, because of these contradicting operation strategies, a complex mode of operation must generally be applied – especially when dealing with multipurpose reservoirs.

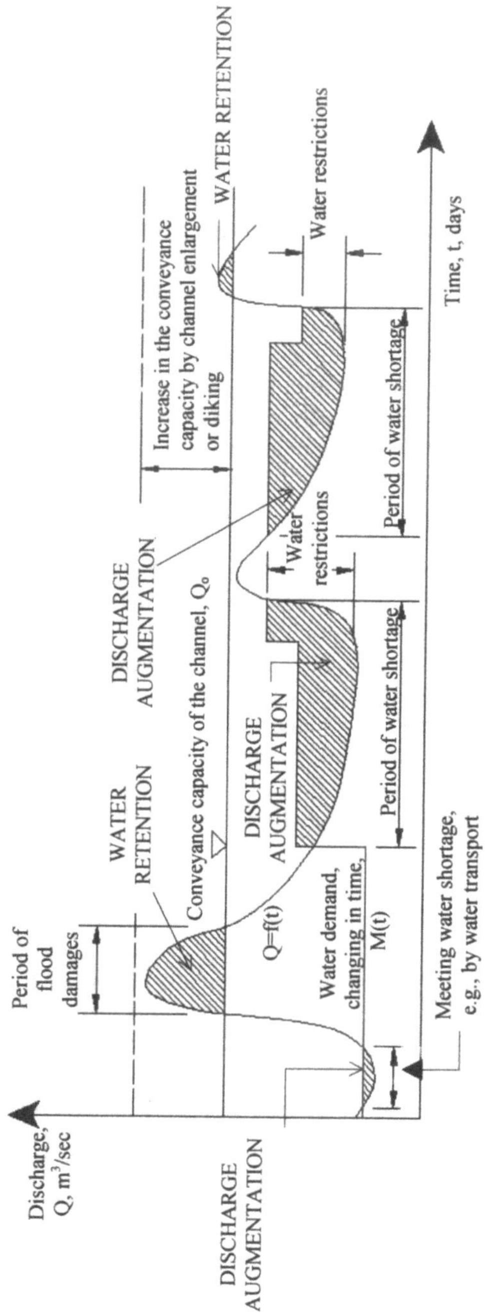

Figure 1.1. **Balancing water supply and water demand by using either natural or regulated flows**

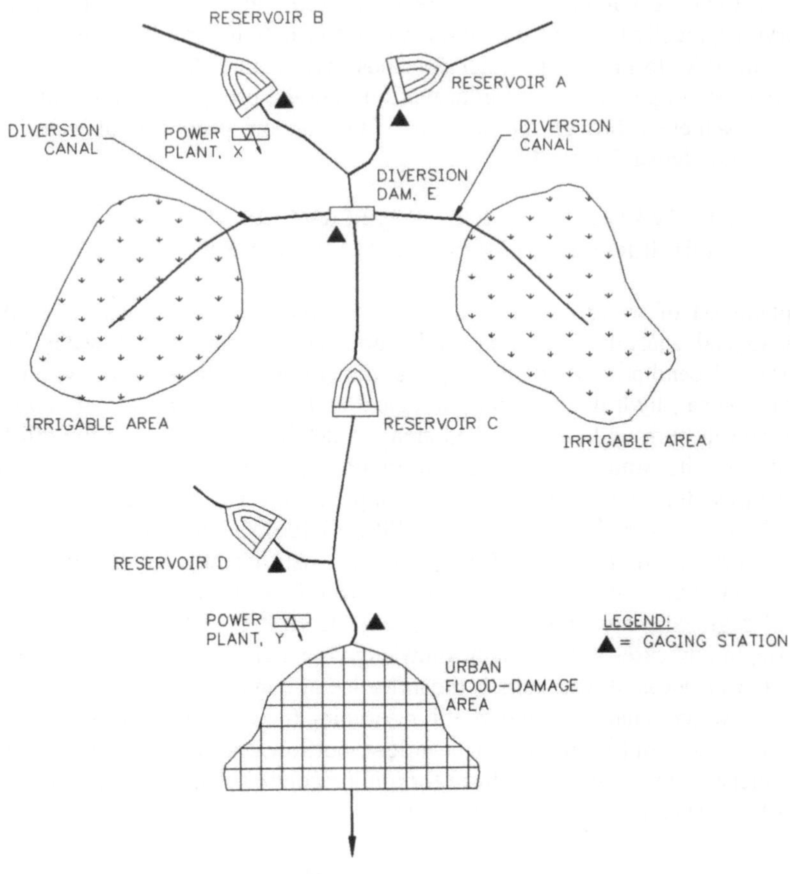

Figure 1.2. **Balancing water supply and water demand by using storage reservoirs**

In reservoirs serving water users only, apart from the assurance of a generally low but guaranteed amount of water withdrawal, the water surplus in wet periods are retained for dry periods. The situation is the same in practice with reservoirs for which distinctly allocated storage spaces serve the different water users in order to meet the flow control needs. The reservoir is kept filled up as needed, while the retention of floods is assured by storage spaces above the conservation needs, and/or kept empty for as much and as long as the designated downstream flooding criteria would permit. These patterns of operation are mutually beneficial since they could affect one another. For example, a volume of water retained from large floods and released from a reservoir typically shortens the duration of water deficit period; and filling up a storage space in order to meet high water demands may require the storage of an entire flood hydrograph (Figure 1.3). For both types of storage problems, numerical solutions can only be achieved through the analysis of the random processes of water supply and, often, water demand as well.

1.2.2. THE WATER BALANCE OF RESERVOIRS AS A TOOL IN RESERVOIR DESIGN AND MANAGEMENT

Computation of water balances is one of the most important aspects in the field of hydrological analysis. The input and output components of the water balance of a reservoir depend not only on the physical dimensions of the water body, but also on the climatological, hydrological, morphological and geological factors affecting the water body and its surrounding watershed area. A detailed discussion of the effects of these factors on the water balance of reservoirs, as well as elaboration of evaluation procedures for the water balance components, are provided elsewhere in the hydrological literature (e.g., UNESCO, 1981; UNESCO, 1974; Bruce and Clarke, 1966; Chow, 1964). In the case of ungauged or insufficiently gauged water bodies, an estimation of the water balance components can, in some cases, be obtained through the use of regional maps and atlases (see, e.g., UNESCO, 1982; NERC, 1975; Klemes, 1973). In any case, it must me acknowledged that an increasingly important aspect of the water balances of reservoirs is the influence of human activities.

The water balance equation for reservoirs for any time interval is a continuity equation, with equilibrium between inflow, outflow and storage components for each time interval. In its simplified form for routine control of water inflow and outflow, the water balance equation can be written as:

$$\sum I - \sum Q = \frac{\Delta S}{\Delta t} \tag{1.1}$$

where: ΣI = total inflows; ΣQ = total outflows; and ΔS = change in water storage in the reservoir in the period of balance, Δt. This represents the classical and basic continuity equation used in the design of reservoirs; evaporation and precipitation terms may also be explicitly specified in this equation.

In general, water balance relationships form the basis for rational deterministic hydrological forecasting models. Quantitative estimates of the available and/or potentially available water resources can be used to determine the optimum usage necessary to satisfy the various economic needs of the project area.

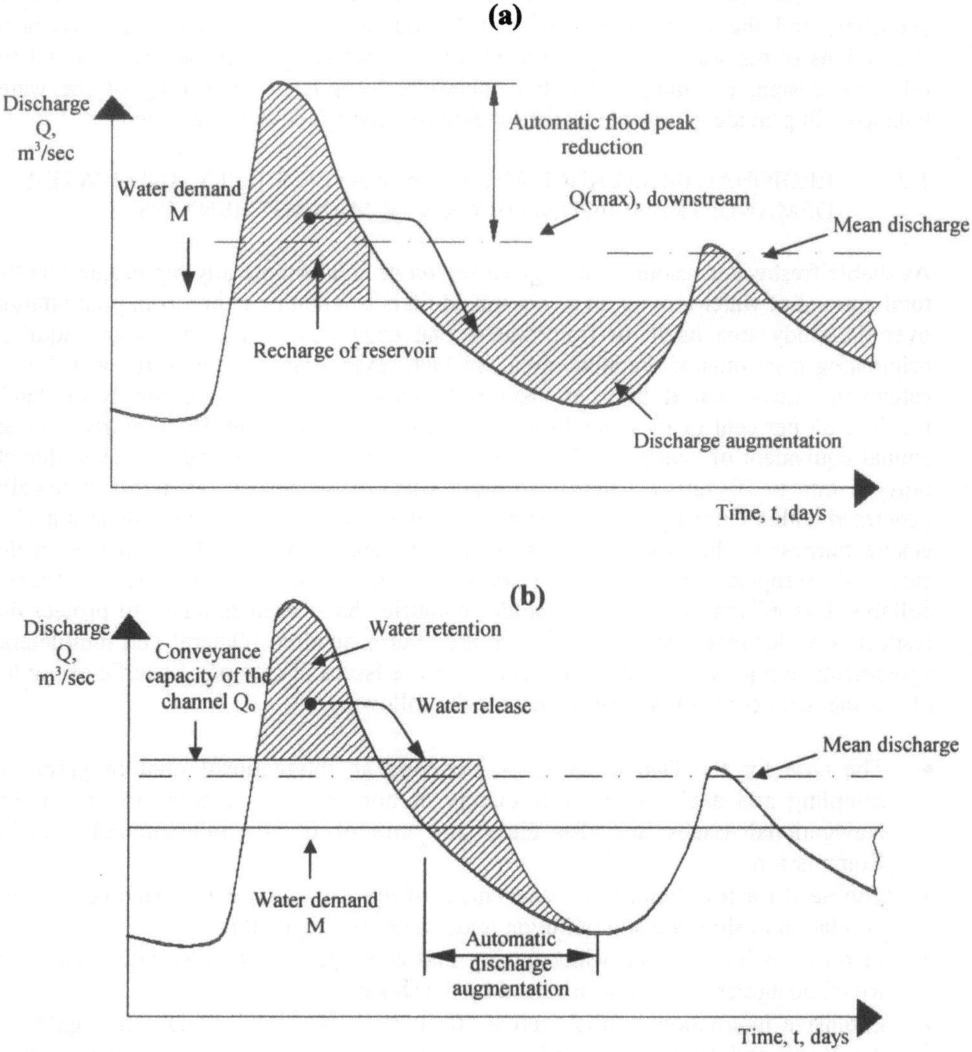

Figure 1.3. Storage of entire flood hydrograph to meet high water demands in the joint operation of reservoirs: (a) for various uses, and (b) for flood control

It is apparent that the water storage in reservoirs serves a variety of human activities, and the optimum use of these resources requires effective management. Predictions of the water balance components of a projected reservoir are essential for effective design, planning and initial operation. Long-term monitoring of the water balance will provide a valuable statistical database for effective management.

1.2.3. REGIONAL IMBALANCE BETWEEN WATER SUPPLY AND WATER DEMAND: TRANSBOUNDARY RIVER MANAGEMENT ISSUES

Available freshwater resources in a given region or area is generally represented by the total amount of water moving in rivers and aquifers originating either from precipitation over the study area itself, or from waters that enter this area in rivers and aquifers originating from outside the study area. In fact, several large rivers of the world have catchment areas shared between many countries. For example, the Netherlands receives 85 per cent of its water from 'transboundary' rivers, and Hungary receives an annual equivalent of nearly 1,200 mm over the country from the Danube, even though only 65 mm to 70 mm (i.e., less than 6 per cent) of the associated runoff is actually generated within Hungary. Consequently, transnational efforts are generally required to ensure fairness in the utilization of such shared water resources. For instance, in the case of European rivers (e.g., Danube, Rhine, Elbe, Neman, Dvina, Maas), collaborative efforts between stakeholder countries have been initiated to protect the respective water resources; indeed, there are over a hundred bilateral and multilateral agreements to regulate water quality and water use issues in Europe. Specific examples of transnational cooperative efforts include the following:

- The case for the Danube Basin – involving an international joint program for sampling and analysis of water quality at border crossing points on the river. Navigational issues are also under the control of the International Danube Commission.
- The need for transboundary management of the Rhine River has been recognized in relation to shipping and pollution issues associated with this river.
- In order to improve the water quality and ecological conditions within the Elbe River, an agreement has been signed by the stakeholder countries.
- Extensive international collaborative efforts have been taking place in regards to the North and Baltic seas, and also the Mediterranean – with a goal to reduce riverine releases of nutrients and dangerous substances.

Despite the cooperative spirit in a number of instances, such as those illuminated above, it should be acknowledged that some water resources, industrial pollution, and/or dam construction plans often can create some level of international tension that leaves a negative conflict-like trail. A positive example of things going well, however, relates to the Columbia River Treaty (of 1964) that followed from a negotiation between the United States and Canada for joint development of the Columbia River. The terms provided for a transboundary storage project, with the dam site situated in the US – and the reservoir headwaters extending into British Columbia, Canada. In that case, the US acquires large quantities of Canadian storage to meet certain power and flood control objectives, and Canada receives a share of the increase in power produced at US generating plants – plus payment for its storage contribution toward flood damage reduction within US borders. The Columbia River Treaty indeed represents an outstanding example of international co-operation in river basin development.

In general, significant regional imbalances between water supply and water demand tends to exist globally. In addition, multi-annual dry or wet periods can occur over large regions, and climate change can also worsen the water resources situation in many regions. The development of effective flow regulation strategies must therefore address the international or cross-boundary perspectives of the flow regulation scheme.

1.3. Impact of Reservoir Projects on the Temporal and Spatial Variation of Streamflow Quantity and Quality

The creation of reservoirs can have significant impacts on the quantity and quality of the resulting water body when compared to the pre-project conditions. The physical changes associated with building dams on complex hydrological systems are indeed numerous (Baker *et al.*, 2000). Some of the very important physical factors that bear on the utility of a reservoir project – in addition to the obvious effects of reservoir storage on downstream flows – relate to evaporation, hydrochemical regime (or environmental quality), groundwater conditions, and sediment movement. Typically, evaporation reduces the quantity and quality of available water, and sediment deposits decrease the amount of available space for water storage. Also, reservoirs may increase the runoff in some situations because precipitation falling directly into the reservoir water surface is not subject to infiltration losses; on the other hand, reservoirs may decrease runoff by increasing the evaporation and groundwater percolation out of the basin. In general, the construction of reservoirs can cause major changes in streamflow volume and quality, as well as its distribution in time and space.

The primary goal of the dam project is to ensure sustainable benefits that also accounts for factors such as water quality, public health, aquatic and terrestrial system sustainability, and the natural cultural value of the area. Although reservoirs are usually built for a primary use, very often its associated secondary problems become an inseparable part of the overall project. Thus, unless both the primary and secondary aspects are thoroughly evaluated in the very early stages of reservoir planning, economic loss and human suffering could be enormous. In fact, it is an indisputable fact that the construction of dams can have profound impacts on a river system; the dam project also tends to affect existing human and non-human lives that are either dependent on the river or simply inhabit areas around the river basin or its watershed.

Indeed, the philosophy for a reservoir design problem should incorporate the major economic, technical and social characteristics of the transformed environment. It is obvious that society cannot accept a design strategy in which, for instance, only the energy production needs is the sole project objective – without consideration to other non-energy but equally important issues. However, if the reservoir will satisfy the needs of flood protection, navigation and transport, recreation, and still meet environmental and aesthetic standards, then the project is justifiable.

1.3.1. ENVIRONMENTAL IMPACT ISSUES ASSOCIATED WITH DAM AND RESERVOIR PROJECTS

Reservoir projects can affect the quality of the aquatic and terrestrial environment – including the preservation of certain wildlife habitat and changes in the local climate.

Construction of large reservoirs may also have significant effects on the local economy due to displacement of people and changes in agricultural (farming, fishing, etc.) techniques. In addition, changes in the ecosystem may increase the incidence of human parasites and diseases. Furthermore, in certain regions of the world where there is seasonal winter snow, reservoir projects will normally change the ice conditions in the catchment. Regulation of the reservoir would invariably cause fluctuation in the water levels, which would possibly lead to the cracking of the ice cover with consequential water bleeding; this can create problems for traffic to and from the ice zone. Common types of environmental impact issues associated with dam and reservoir projects are annotated below.

- *Water-Borne Diseases.* One of the classic health challenges associated with reservoir projects involves malaria, the mosquito-borne disease that tends to affect the population/settlement near some of the swampy areas along the original river. An appropriate program should therefore be developed that controls the mosquito vector – with emphasis placed on providing an ecological environment that is unsuitable for its propagation. In fact, even in non-tropical or similar environments where malaria may not exist, similar consideration is needed for all other water-related diseases (such as for river blindness and bilharzias) in the project area. In particular, preventive measures are required to combat/prevent increase in such water-related diseases around head-pond areas.

- *Water Quality Parameters.* A common problem area in water impoundment situations is that of low levels of dissolved oxygen (DO) in the releases from deep reservoirs. This has been an especially important issue because many deep impoundments have been built without providing ways to withdraw water selectively from the reservoir strata that have the higher DO content. Other water quality parameters affected include water temperature, biochemical oxygen demand, turbidity, and indeed a variety of other chemical and biological properties.

- *Eutrophication.* A major concern of water quality management that should not be overlooked is the issue of eutrophication – i.e., the nutrient enrichment of the water and the bottom of reservoirs and lakes. Indeed, human-made eutrophication has been considered one of the most serious global water quality problems of lakes and reservoirs during the past two to three decades. Increasing discharges of domestic and industrial wastewater, the intensive use of crop fertilizers, the rise in airborne pollution and the natural mineralization of the inflow can be seen as some of the major causes of this undesirable phenomenon. Typical symptoms of eutrophication include, among others, sudden algal blooms, water coloration, floating water-plants and debris, and excretion of toxic substances that causes taste and odor problems in the production of drinking water, and sometimes fish kills. These symptoms can result in limitations of water use for domestic, agricultural, industrial or recreational purposes. Consequently, eutrophication can adversely impact the region where the storage reservoir has been placed – i.e., in the absence of an appropriate water quality management policy.

- *Stream Morphology Changes.* River flow regulation generally will result in the natural balance between rivers and their floodplains being disturbed. Hence, it is crucial to develop a comprehensive plan of action to deal with potential problems resulting from any catchment modifications. Examples of the effects of the disturbance may be that, as a result of the flow regulation, the heterogeneous meandering river may change into a homogeneous straight channel with less habitat diversity, increased transport of sediments, and reduced self-purification.

In general, the construction of storage reservoirs will typically cause a change in the conditions of surface and subsurface runoff, environmental characteristics, and prior existing economic activities. Consequently there is a need to carefully analyze the whole complex set of issues in a comprehensive impact assessment that addresses both the quality and quantity aspects of the flow regulation.

Overall, it should always be the objective of development projects to increase the total quality of life of the people in the project area in a long-term sustainable manner. Indeed, the many considerations and factors involving both technical and economic components of a project can be balanced with the environmental and sociological values – but this requires creative thinking by all concerned parties or project analysts. With a scheme that ensures the early identification of the environmental benefits and their costs, together with additional economic costs and benefits, a generally more acceptable project can be implemented. Of course, the environmental benefits are usually difficult to evaluate in monetary terms; consequently, effective novel strategies for achieving consensus often become particularly important. For instance, some benefits that cannot be quantified must be evaluated qualitatively. Ultimately, however, the overall process can be designed to produce a well-balanced and sustainable project that is acceptable to all stakeholders.

1.4. Using Storage Reservoirs in Flow Regulation and Water Management Schemes

Storage reservoirs represent one of the most effective tools for eliminating discrepancies in the time and space variations of water resources distribution and requirements. In fact, the different – often contradictory – and increasing demands on water resources utilization and control usually can be fulfilled only by building multi-purpose reservoir systems. In this way, the available water resources can be exploited and/or managed in a more rational way.

Typically, the construction of a dam across a river valley causes water to accumulate in a reservoir behind the dam; the volume of water accumulated in the reservoir will depend, in part, on the dimensions of the dam. The size of the dam will normally affect the capital expenditure in a very significant way. Indeed, the construction of large water resource control systems – such as dams – generally involves rather huge manpower and material outlays. Consequently, the elaboration of effectual methods of approach that can be used in establishing the optimal reservoir parameters is of great practical significance. For instance, in the design and operation of large multi-reservoir systems, simple simulation and/or optimization models that can identify potentially cost-effective and efficient system design are highly desirable. But it should also be recognized that the problem of finding optimal capacities for multi-reservoir systems (particularly for a system in which reservoirs are acting both in parallel and in series, and/or dependently and independently) often becomes computationally complex because of the large number of feasible configurations that usually need to be analyzed. This calls for the use of contemporary techniques and tools – especially for project planning; reservoir site selection; water resource systems operation; best management practice decisions; water quantity and quality assessments; monitoring studies; flood forecasting; flood control assessments; etc. This book elaborates some systematic methods of approach to the hydrological design and operation of reservoir storage capacities that can be employed to help optimize the reservoir design process. The application of such novel techniques to hydrological problems are indeed very important – especially because problems often encountered in hydrological designs tend to require repeated sampling of the variables of interest – in order to obtain information over large areas, and also for the integration of all such data. In addition, these methods of approach can allow a water resources system designer or analyst to engage in a variety of scenario evaluations that satisfy different project conditions – which can therefore facilitate an optimization of the overall system design and operation. By using the systematic approach offered by this title in a reservoir system design effort, different policies can be evaluated and compared, which will then facilitate the selection of an optimal design and operation plan for a proposed or existing dam project.

Overall, this book attempts to provide a concise, yet comprehensive overview of the hydrological dimensioning aspects relating to the design and operation of river dams. It presents some very important concepts, tools and methodologies that can be used to help resolve multiple reservoir capacity allocation problems in a consistent, efficient and cost-effective way. The material presented in the book will guide the hydrological analyst or design engineer in the determination of the optimal storage capacities and operational strategies of the individual elements in a system of reservoirs present at a set of potential reservoir dam sites. The specific goal in the type of optimization problem elaborated here will be to help minimize the reservoir capacities and/or shortfalls (which are a surrogate for minimizing costs) and still meet capacity requirements and needs. The ultimate objective in such application will be to prevent over-design or under-design of reservoir storage capacities under optimum conditions of reservoir operation. Thus, the optimal solution will produce the smallest reservoir storage required to meet all project demands with an acceptable degree of reliability.

In the reservoir design process, the underlying design concept pre-supposes that, when dealing with a multi-reservoir system, the individual reservoirs are the basic design elements and that their water resources are utilized for some defined multi-purpose use(s). The methods presented in this book will also help answer the following important capacity allocation question: 'what fraction of the total demand required for a delineated supply area should come from each reservoir, *i*, acting jointly with all the other reservoirs in the system of *n* reservoirs?' This then forms the basis for the allocation of storage capacities to the various reservoir sites that will be necessary to meet the demands with a desirable reliability. Of course, when only one site is involved, then the problem reduces to allocating the total storage to the single reservoir.

Finally, it is noteworthy that, notwithstanding computers and mathematical optimization procedures, reservoir design can be viewed as a creative art – and the quality of design depends substantially on the designer's ability to visualize the interaction of all components of the proposed system.

Chapter 2

THE HYDROLOGY OF FLOW REGULATION

Flow regulation is a consciously performed re-distribution of runoff – often achieved by use of human-controlled hydraulic control structures such as storage reservoirs (see illustration in Figure 2.1). There are two basic types of flow regulation schemes – depending on whether the purpose of the control structure is to stabilize some lower limit of outflow (i.e., low-flow regulation) or some upper limit to maintain a certain degree of flood protection (i.e., flood control). In flow regulation schemes using storage reservoirs, that portion of the reservoir volume that can be used for the flow regulation is referred to as the reservoir storage capacity.

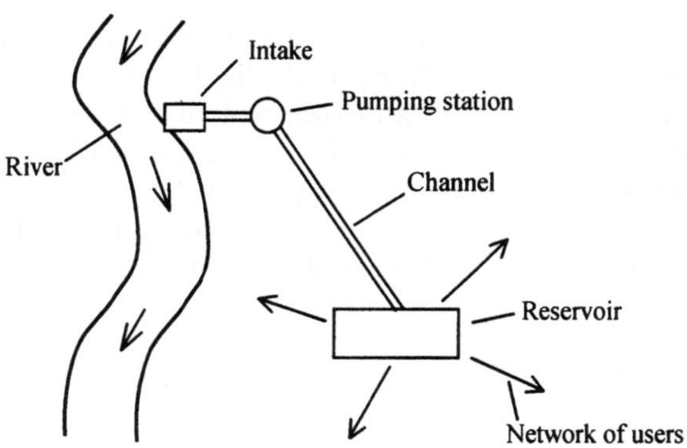

Figure 2.1. Schematic of typical structures for flow regulation

2.1. Flow Allocations

A reservoir may typically be required to provide primarily for hydropower generation, irrigation, water supply, flood control, navigational and recreational uses, etc. Any one of these requirements of the reservoir – many of which place contradicting storage needs of the reservoir storage capacity – imposes certain constraints on the operating policy; for instance, power generation requires a full reservoir or maximum head whereas flood protection requires an empty reservoir. In any case, the allocated flow $q_i(t)$ to a particular use at site *i* in period *t* must be no greater than the total flow, $Q_i(t)$, available at that site and in that period, viz.:

$$q_i(t) \leq Q_i(t) \tag{2.1}$$

The quantity of water that any particular user expects to receive in each particular period is called the *target allocation.* Given an known or unknown annual target allocation at site *i*, T_i, some – usually known – fraction $\alpha_i(t)$ of that annual target allocation will be needed in period *t*. Consequently,

$$q_i(t) < \alpha_i(t)\cdot T_i \implies \text{a deficit, } d_i(t)\text{, results} \tag{2.2}$$

$$q_i(t) > \alpha_i(t)\cdot T_i \implies \text{an excess, } e_i(t)\text{, results} \tag{2.3}$$

Hence,

$$q_i(t) = [\alpha_i(t)\cdot T_i\,] - d_i(t) + e_i(t) \tag{2.4}$$

where either $d_i(t)$ or $e_i(t)$ or both are zero. The dependability that outflow will neither be lower than draft nor higher than the flood limit can be measured by an index that defines the reliability of the flow regulation scheme.

Whether or not any deficit or excess allocation should be allowed at site *i* depends on the quantity of water available and the losses associated with deficit or excess allocations to that site. At sites where the benefits derived in each period are independent of the allocations in other periods, the losses associated with deficits, and losses or benefits associated with excesses can be defined in each period *t*. The average flow just downstream of the withdrawal site, $Q_{i2}(t)$, will equal the original flow at that site, $Q_i(t)$, less the quantity withdrawn, $q_i(t)$, plus any 'return flow' fraction, $\rho_{s,\tau,t}$, of the amount withdrawn in the current and previous period *t*. Assuming all the return flows enter just downstream of the point of withdrawal, then

$$Q_{i2}(t) = Q_i(t) - q_i(t) + \sum_{\tau<t}^{t}\left(\rho_{s,\tau,t} \times q_i(\tau)\right) \tag{2.5}$$

where $\rho_{s,\tau,t}$ is the return flows as a fraction of the amount withdrawn in the current and previous period up to time period *t*; and $q_i(\tau)$ is quantity withdrawn in the period τ. Water stored in reservoirs can often be used to augment downstream flows for in-stream uses such as recreation, navigation, and water quality control.

2.2. Hydrological Problems Resulting from Water Deficiency

The notion of water deficiency is closely related to that of drought – albeit there is profound difference between the two. Klemes (1973) provides a comprehensive discussion that distinguishes between water deficiency and drought.

Water deficiency arises if, from the point of view of human requirements, there is not enough water available; the difference between water available at a given location and time, and the concurrent water demands (including losses), represent the actual deficit that has to be eliminated by water from some geographical location and/or storage accumulated during periods of water surplus.

The basic hydrological problems of low-flow regulation is to find a mathematical relationship between reservoir storage (S), draft or water demand (D), and dependability or reliability (R) – so that the storage equation is described by the following functional relationship:

$$S = f_S(D, R) \tag{2.6}$$

or, by the following alternative formulations of the same relationship:

$$D = f_D(R, S) \qquad \text{or} \qquad R = f_R(D, S) \tag{2.7}$$

Indeed, the solution of the above equations is dependent on the properties of reservoir inflow (Q) and on the nature of release policy adopted.

In the deterministic methods of solving the storage equation, the reservoir input parameters are derived solely from historical streamflow records. The solution of storage equation obtained on this basis is thus valid for this past period, and any extrapolation in time could prove unreliable. Nevertheless, depending on the reliability of the forecasts, it has been standard practice in water resources management to design storage reservoirs on this basis; the solution can be found either numerically or by means of graphical techniques.

Stochastic methods of solution to the storage equation usually utilize simulation techniques that involve the generation of synthetic streamflow series for use as surrogate to the historical flow record. Other stochastic methods consist of the analytical-stochastic and numerical-stochastic techniques.

Hybrid methods of solution to the storage equation – that consist of part-stochastic-part-deterministic (or combinations to different degrees of stochastic and deterministic techniques) procedures – offer yet other opportunities for improving the evaluation schemes.

2.3. Hydrological Problems Caused by Water Excess

The presence of more water than is desirable at a given location, from the point of view of human or society needs, gives rise to the notion of water excess. Of particular interest in reservoir hydrology is the water excess that is temporary and is a consequence of an extreme hydrological situation – i.e., a flood; this type of water excess or flood is often catastrophic in nature and tends to have limited time predictability and/or sudden occurrence.

A flood is that flow of a stream or river that causes an overflow/overtopping of its natural or artificial banks. From the water management point of view, the main feature of a flood is the degree of its interference with human activities – measured by the extent of actual or potential economic loss and/or danger to human lives. For a given river reach with a given type of land use along the river, the harmfulness of a flood generally depends on its magnitude. The flood magnitude may relate to several physical features of the flood wave – including, flood elevation, flood discharge, flood volume, flood duration, seasonality in the flood occurrence, and flow velocity; these important terminologies that find common usage in flood control studies are annotated below.

- *Flood elevation* is the maximum elevation of water level encountered during the flood; it is not always the best indicator of flood harmfulness.

- *Flood discharge* is the maximum rate of flow during the flood; the peak discharge of a flood usually closely coincides with the occurrence of maximum water level in the river, thus being a good indication of flood elevation. In reality, the peak discharge occurs before the maximum water level is reached but in most instances the difference between the maximum water level and that corresponding to the peak discharge is not significant.

- *Flood volume* is the volume of water in a flood wave above a given discharge limit, Q_i – defined as follows:

$$V_i = \int_{Q_{max}}^{Q_i} t dQ \tag{2.8}$$

  This can easily be obtained from a flood hydrograph (see Figure 2.2, as an example) and represents an indispensable characteristic wherever flood damage depends on the quantity of flooding water, or in cases of flood control aimed at a reduction of flood peak. From the point of view of flood protection, the flood volume is by far the most important flood characteristic.

- *Flood duration* is the period of time during which discharge does not drop below a safe limit. Quite often this characteristic is closely related to the danger the flood poses.

The *flood-frequency curve* is used to show the average interval time (or recurrence interval) within which a flood of a given magnitude will be equaled or exceeded in any given year. It is noteworthy that, the *seasonal occurrence* of floods could have beneficial or detrimental effects on land use, depending on when and where it occurs; for instance the fertilizing effect of suspended sediments and minerals carried in flood waters could be used to an advantage if such flood occurs at the 'proper' time in the 'right' place. Also, *flow velocity* during flood events may be a determinant of the extent of damage to engineering/hydraulic structures, as well as affect other water uses and further cause undesirable changes in river channel morphology by wave erosion.

The ultimate objective of flood analysis is to assess the degree and frequency of interference of floods with normal life and activities of a community in order that adequate preventive measures can be taken. It is usually assumed that most adverse effects of floods can be reasonably related to their peak discharges; thus, the main objective of flood analysis reduces to finding the frequency of occurrence of a flood with a given peak discharge.

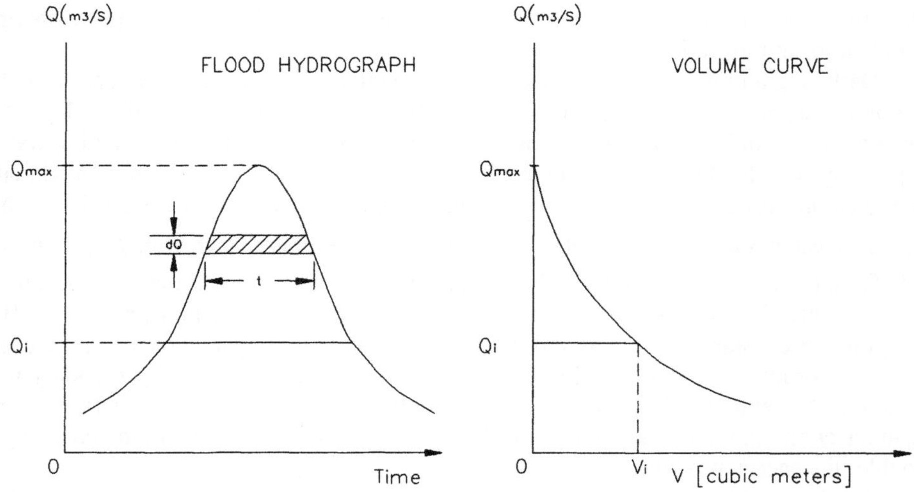

Figure 2.2. Definition sketch for the derivation of the flood volume curve

2.4. Types of Flow Regulation

Depending on the duration considered in a flow regulation scheme, distinction can be made between the following principal types of flow regulation: daily, weekly, seasonal (or annual) and multi-seasonal (or multi-year). These general and broad types of flow regulation are briefly elaborated below. Of course, other types of flow regulation schemes can be developed for case-specific situations – as, for instance, in the case of non-periodic short-duration schemes. The essence of non-periodical, short-duration flow regulation is that the stored water from the reservoir is carried off in a non-periodic manner. This type of regulation may occur in cases when, for instance, there are needs to satisfy requirements for the navigation of ships or for the passing of ice through a watercourse having limited depth in the low flow period. Another example relates to water supply for irrigation systems, when the storage volume of the reservoir is limited, and there is a possibility to provide supplies to the system during one or two days in the low flow period. This type of flow regulation is typically used where small rivers are involved.

2.4.1. DAILY FLOW REGULATION

The typical objective in studies involving short-duration flow regulations is the transformation of continuous flow information into non-continuous flow regimes – as demonstrable by daily flow regulation schemes.

Daily regulation schemes allows for a possibility to increase the number of users in such a manner that water volumes accumulated during the hours of reduced water supply is used to satisfy the subsequent peak needs of the users. The benefit of such type of regulation is the ability to lower the construction costs of appurtenant hydraulic structures and intake works, especially in the case of drinking water supply, energy production, and irrigation.

Daily regulation schemes typically find specific applications in situations with variable (i.e., non-constant) demands by the users during different hours of the day. An example of the daily regulation of flows necessary to meet demands of a drinking water supply system is illustrated in Figure 2.3. Here, $q_{M,max}$ represents the maximum demands and $q_{M,min}$ is the minimum demands during a day. If it is assumed that the needs of water supply can be satisfied by a constant average discharge, $\overline{q}_M$, then the satisfaction of the maximum needs, $q_{M,max}$, in the mid-hours of the day becomes possible only if the volumes of water, W_2 and W_3, have been stored in the reservoir. By utilizing these volumes, the required volume, W_1, can then be guaranteed. In the case of power production, it is possible to increase the energy produced by reducing the duration of operation. For example, if the inflow is stored in the reservoir for a prior 12-hour duration, then in the latter part of the day, it becomes possible to increase (e.g., double) the energy produced.

For the drinking water supply, the pumping station and the main channel can be calibrated for the discharge, q_{av} – and from the daily regulation scheme for the reservoir, we are able to satisfy the q_{max} needs.

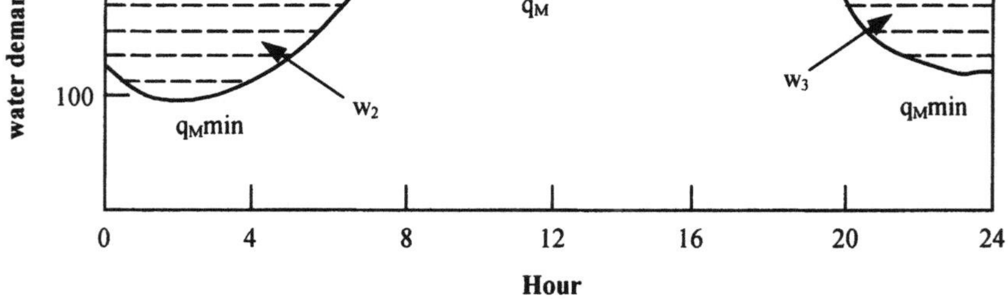

Figure 2.3.
Illustration for a daily flow regulation scheme

2.4.2. WEEKLY FLOW REGULATION

Weekly flow regulation schemes are most applicable in situations such as exist during weekends and holidays – i.e., when major consumers (e.g., industrial units, institutions, etc.) are typically not operating, and consequently the demands for water supply and energy production is lower than normal (Figure 2.4). Here, $q_{1,M}$ is the water demand during the working days, $q_{2,M}$ is the demand for the holidays, and $\overline{q}_M$ is the average value for the number of the days, *n*. If (n-2) is the number of working days (i.e., assuming two holidays in a week), then:

$$\overline{q}_M = \frac{q_{1M}(n-2)+2q_{2M}}{n} \tag{2.9}$$

Consequently, it is possible to decrease the calculated demand Δq_M, viz.:

$$\Delta q_M = q_{1,M} - \overline{q_M} \tag{2.10}$$

Substituting this value from equation (2.9) into the expression in (2.10) yields the following:

$$\Delta q_M = q_{1,M} - \frac{q_{1M}(n-2)+2q_{2M}}{n} = \frac{2(q_{1M}-q_{2M})}{n} \tag{2.11}$$

From this relationship, the weekly capacity requirements of the reservoir can be determined and balanced in consideration of the volumes W_1 and W_2 indicated on Figure 2.4.

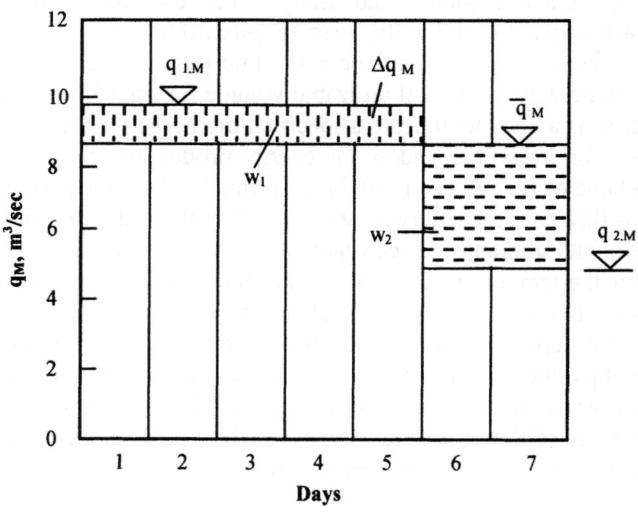

Figure 2.4. Illustration for a weekly flow regulation scheme

2.4.3. SEASONAL OR ANNUAL FLOW REGULATION

The need for seasonal or annual flow regulation is a very common occurrence in water management practice. For such cases, water from the high flow period is stored in the reservoir – and will be discharged in the critical low flow period. The necessary capacity requirements of the reservoir can be determined on the basis of the time series of flow and the water demands. Common examples of this type of flow regulation involve the continuous supply (Figure 2.5) and the non-uniform, intermittent supply (Figure 2.6) scenarios. Here, Q is the natural discharge of the river, and q_M is the demand for meeting water supply requirements.

Reservoirs for seasonal flow regulation can also be used for flood protection purposes when there are means to retain the water volume of the flood waves in order to enhance the flood-protection capacity and ensure safety of the downstream areas. In case of barrage reservoirs, and considering a longer period, this action can be seen as the conversion of the time-function of the upstream discharges into the time-function of the downstream ones. Of course, this conversion depends on the geometric and hydraulic features of the pool, as well as on the reservoir operation strategies adopted.

2.4.4. MULTIANNUAL FLOW REGULATION

The objective of multiannual flow regulation is the storage of discharges of the wet years to such an extent that the stored volumes can satisfy the shortages of the dry years. This method of flow regulation may offer a more 'real-time' control procedure – although it typically requires the construction of rather large reservoirs. Under such circumstances, only a given part of the serviceable volume of the reservoir will tend to be used for the annual depletion and filling of the reservoir. The complete depletion may take place only at the end of dry years (Figure 2.7).

When reservoirs are used for flood control purposes, there is a need for designating a specific volume within the total utilizable volume for transformation of the maximum floods. If there is a need to allocate a specified discharge for drinking and/or industrial water supply, for irrigation, and/or for energy production, then the depletion of the utilizable volume of the reservoir will be performed only in the period of time when the discharges of flow into the reservoir are lower than the water demands. Consequently, the reservoir will guarantee a constant water supply through the years – with a possibility for the temporary (i.e., non-permanent) supply as well during certain years. Any excesses above the needs of users, after filling the reservoir, will be discharged, or in the case of a depleted reservoir, the inflow is used for the filling of the reservoir. Thus, if our objective is a 'non-permanent' water supply, then the reservoir will satisfy not only the given needs of users, but it will also offer a possibility of using overflows in certain periods of time for other purposes. This type of flow regulation is usually employed in the case of reservoirs operating in a system for power production.

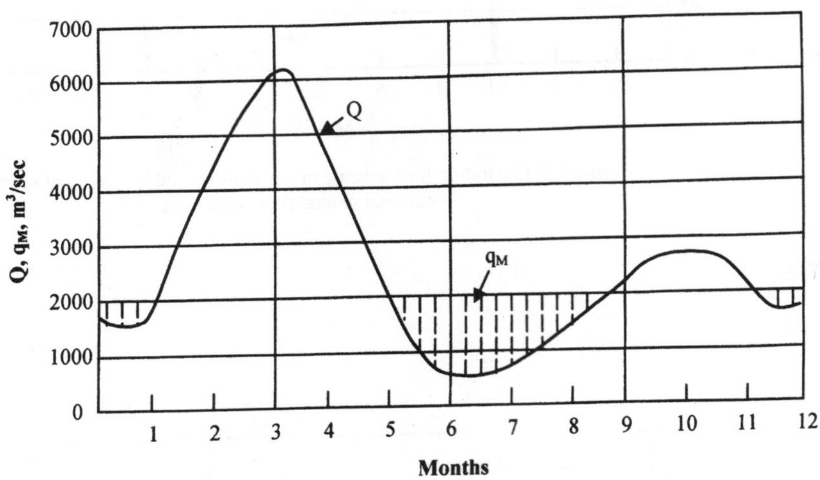

Figure 2.5. Illustration for a scheme of continuous supply – seasonal-annual flow regulation

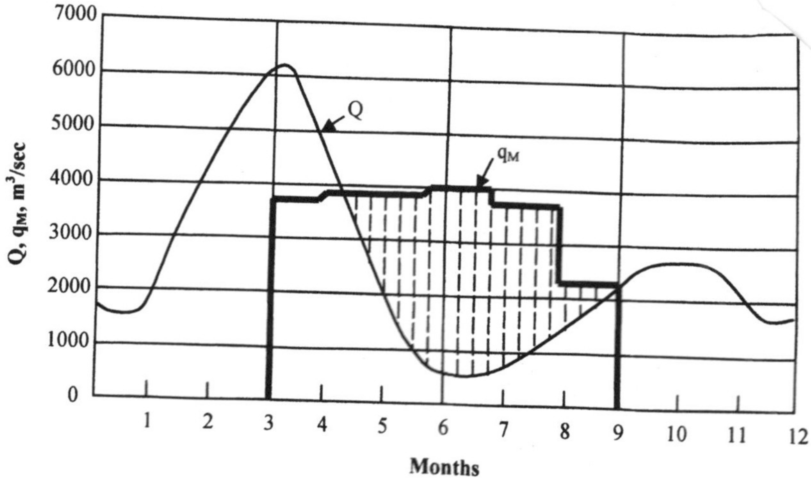

Figure 2.6. Illustration for a scheme of non-uniform and intermittent supply – seasonal-annual flow regulation

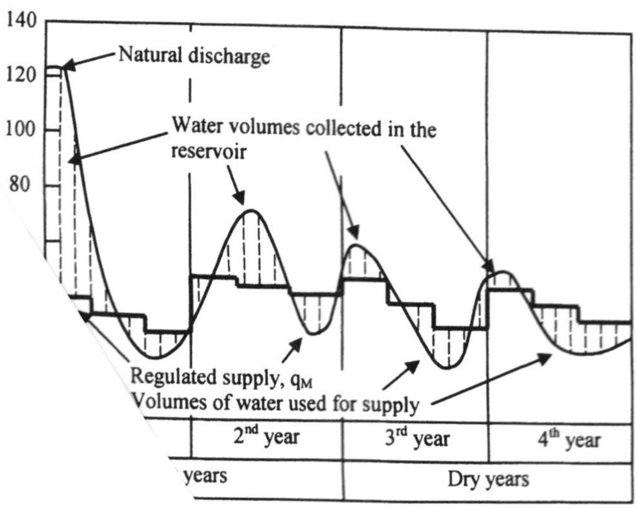

ıstration for a scheme of multiannual flow regulation

Optimizing Flow Regulation Schemes

become the general strategy of choice to regulate flow regimes by careful ration of specific human and/or societal needs – instead of the simplistic :h of accommodation to the flow regime, together with and the passive or ive protection against the damages of floods or drought. From the hydrologic f view, this approach of choice assumes that specific releases can be carried the downstream sections of a river without causing damage to the human or :al environment (Figure 2.8). Typically, however, the flood flows are usually then the conveyance capacity of the riverbed, and the low flows can satisfy the nly within certain time constraints, and even so, at low reliability.

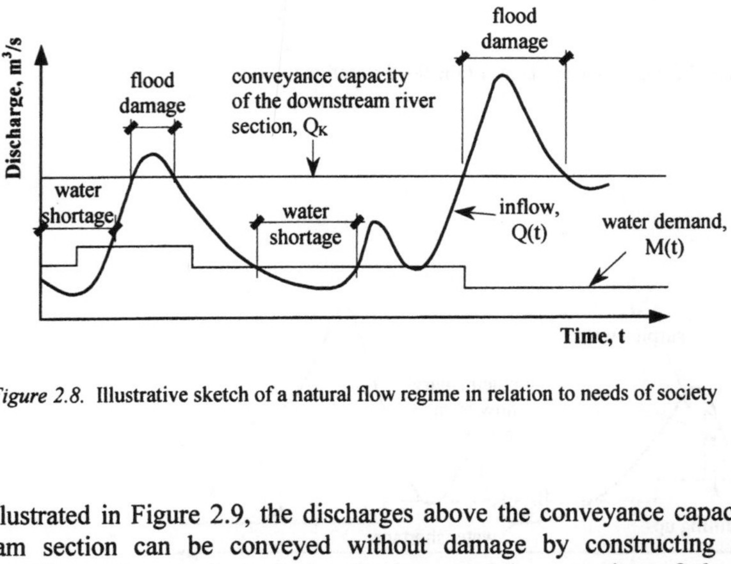

Figure 2.8. Illustrative sketch of a natural flow regime in relation to needs of society

illustrated in Figure 2.9, the discharges above the conveyance capacity of the ream section can be conveyed without damage by constructing a storage ir. The water supply can also be increased by retention of the 'surplus' ges (Figure 2.10).

General Characteristics of Flow Regulation Systems

onstruction of a dam across a river valley causes water to accumulate in a oir behind the dam; the volume of water accumulated in the reservoir will depend dimensions of the dam and the topography of the river valley upstream of the ite. The flow of water from the reservoir will depend on the type of control ures – such as overflow spillways, release gates, sluices, valves, etc. – that are died in the dam design; the flow of water through these structures will depend on effective hydraulic head, which in turn is a function of the elevation of the water in the reservoir.

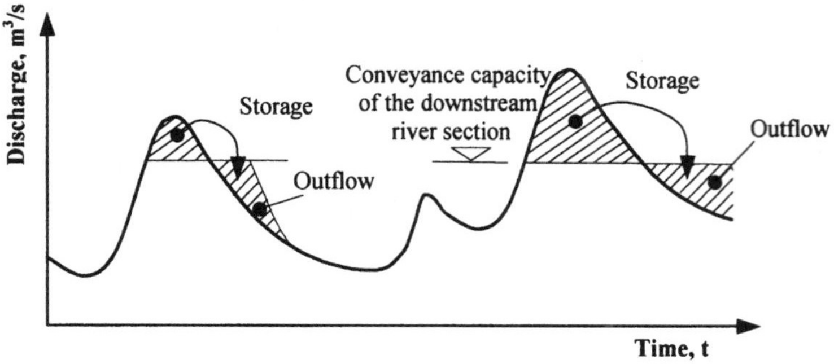

Figure 2.9. Operational scheme of the flood retention reservoir

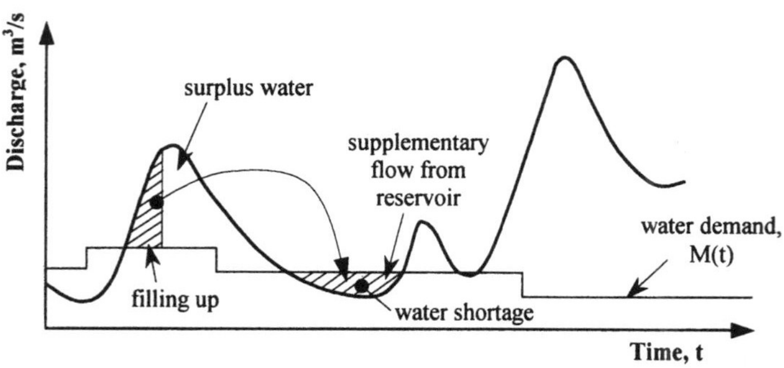

Figure 2.10. Operational scheme of reservoirs for water supply

Storage reservoirs are indeed one of the most effective tools for eliminating discrepancies in the time and space variations of the water resources distribution and requirements. The role of hydrology in reservoir regulation comes into play during the early design of a project and continues throughout its actual operation. Principally, the design of a reservoir requires knowledge of the quantity of streamflow and its occurrence with respect to area and time; and the operation of a reservoir requires the analysis of streamflow based on pre-reservoir flow records, and on current streamflow and precipitation estimates. In general, the basic hydro-meteorological observations necessary for the design, and later on for the management/operation of reservoirs include the following:

1. Precipitation across the watershed – its distribution in space and time (viz., amount, duration and frequency);
2. Evapotranspiration;
3. Infiltration losses/characteristics;
4. River and/or reservoir stages and discharges;
5. Sediment load – including concentration and/or composition of bed load and suspended load.

These observations are vital for the initial planning of reservoirs; they should, nevertheless, be continued during the whole life of the reservoir. Data should be gathered from appropriate points of the entire watershed, pooled together, effectively analyzed, and the results rigorously applied to assure the best use of the available water resources.

It is interesting to note that, depending on the hydraulic conductivity of underlying layers, a human-made lake or reservoir may create and/or increase groundwater levels under certain areas surrounding the reservoir via infiltration into actual or potential aquifers underlying the inundated area. Such newly created or augmented groundwater can be developed for full or supplementary irrigation, water supply or other uses on the lands bordering the reservoir or even at a considerable distance away.

Chapter 3

PLANNING FOR DAMS AND RESERVOIRS: HYDROLOGIC DESIGN ELEMENTS AND OPERATIONAL CHARACTERISTICS OF STORAGE RESERVOIRS

In any region where a new storage reservoir is being proposed, the area must be carefully surveyed to establish suitable sites for the construction of the dam. Each alternative site will then receive a detailed assessment to determine the size of dam that can be constructed and the corresponding storage-area-elevation relationships. With a choice of dam sites and various reservoir sizes, the hydrologist or analyst must now assess the water yield from each site and the magnitude and frequency of extreme floods and low flows. This information will facilitate the selection of the best site to satisfy the water demands of the region, as well as provide flow criteria for the design of the outflow structures and the assessment of compensation water releases required for low flow conditions. In addition, where other reservoirs already exist within the river system/network, an assessment is required of the effect of the new reservoir on inflows and outflows of these existing reservoirs. Invariably, the science of hydrology plays crucial roles throughout these exercises.

Hydrology is the basic science underlying water resources management since water cannot be managed without knowing the quantity, areal/spatial distribution, and the flow dynamics or the timing of the movements of water. One of the most important aspects in the practical application of hydrology to water resources management issues is the computation of basic elements of the hydrological regime – including characteristics of streamflow, precipitation, evaporation, dynamics of the water masses, sediment transport and discharge, water quality, etc. – all of which are essential for the effective design, construction, and operation of water resources projects and hydraulic structures. Safety, costs, and efficiency of projects and structures greatly depend on the reliability of hydrological estimates. But reliable determination of hydrological estimates tends to be complicated – in view of the need to account for possible time series variations for different structures with variable project lives. In any case, hydrometeorological data obtainable from networks form the basis for hydrological computations for most water resources projects. Thus, the quality of such computations depends on the availability of hydrological and meteorological data, their resolution in time and space, as well as their accuracy. Indeed, there are several hydrologic estimates and/or procedural steps of a general nature involved in the planning and design of reservoir projects; UNESCO (1982) presents and discusses such general steps, together with the necessary complementary steps that may be specified in relation to specific projects such as flood control, hydropower and conservation storage. This chapter provides a broad overview of the principal hydrologic and related elements that are important to the design and operation of reservoirs to meet specific project goals.

3.1. The Reservoir Design Problem

Several variables require assessment in the design of reservoirs – the number of variables depending on the type of reservoir. Indeed, the examination of a possible site for the construction of a dam and reservoir requires investigations of a variety of issues. Initially, efforts are coordinated to determine whether construction of a dam and appurtenant features is both technically and economically feasible by giving consideration to the geology of the dam and reservoir sites; the kinds and locations of materials available for construction of the dam; and economic, hydrologic, human and geographical factors. Thorough investigation by boreholes, and geophysical exploratory surveys of the types and distribution of material below the land surface at site(s) of the proposed dam(s) and reservoir(s) is a necessary prerequisite to the design and construction phases of the dam project. Also, in the reservoir planning process, it becomes necessary to conduct a mapping and classification of vegetation of the locale, and also an evaluation of environmental impacts, together with the preparation of multidimensional models of evolution of vegetation. Furthermore, in some cases, it is important to take the problems related to site archaeology and paleontology into account. All such works need to be planned well ahead so that there is no interference with construction phase of the program implementation.

Every reservoir would, in general, be characterized by a number of important and unique relationships – namely the volume-elevation, area-elevation, and elevation-discharge relationships. The knowledge of surface area is important in estimating evaporation loss from the water surface. The elevation of the water surface and the uncontrolled discharge of water from the reservoir are dependent on the spillway structure. In addition to the uncontrolled discharge characteristics, a reservoir will have controlled discharge rules, depending on its function. In the design of new reservoirs, a primary goal is to select the best combination of physical characteristics.

The principal basic data needed for reservoir design studies include adequate topographic maps and hydrological records. Records of streamflow are essential for determining the amount of water available for conservation purposes. Historical streamflow data is better supplemented with stochastically generated flow to provide a more reliable design. Rainfall records are also used to supplement streamflow records, or as a basis for computing streamflow where there are no flow records. Although structural-design studies and social problems arising from population adjustments in a locality are not directly related to hydrology, these should, nevertheless, complement the overall planning scheme. It is also noteworthy that, every reservoir that impounds water behind a dam is a real or potential threat to those living downstream of such structure; also, in some locations where earthquake shocks, movements along bedrock faults beneath dams, or collapse of large volumes of earth materials into the reservoir are distinct possibilities, even the most skilled design and continued maintenance may not preclude failures that could be disastrous to life and property. Where the possibility of massive failure in slopes of reservoirs exist – whether empty or full – appropriate steps should be taken to stabilize the slopes; the particular remedy employed depends on the nature of local conditions. Furthermore, in the case of riverflow regulation, the natural balance between rivers and their floodplains can be disturbed; thus, it is necessary to elaborate a comprehensive plan of action in relation to management of the catchment area. In fact, as a result of flow regulation, the heterogeneous meandering river may at times change into a homogeneous straight channel with less habitat diversity, increased transport of sediments, and reduced self-purification. Such potential problems should be anticipated, and feasible corrective measures incorporated into the overall design scheme.

In general, reservoir size selection usually follows after site selection, but the two may be interdependent. The factors pertinent to size selection would normally include the project function(s), the physical factors, economic factors, environmental considerations as well as social considerations. For economic reasons, it is important that the required water-storage capacity be adequate but no more than necessary. In any case, reservoir operation usually offers significant flexibility – so that optimal conditions can be achieved by making appropriate post-construction adjustments to the reservoir system.

3.1.1. RESERVOIR-SITE SELECTION

The choice of a dam site is governed by the purpose(s) for which the water resource(s) is to be developed; the physical suitability of available sites to serve those purposes safely and economically; and the necessary permission or government authorization to use the particular location of choice. In any case, no two river basins are completely alike – differing not only from the engineering viewpoint but also in cultural, social and economic background and are, in some cases, affected by local, national or even international political issues. Thus, apart from being faced with engineering problems, river basin development can involve several issues and considerations that are entirely non-technical in nature – and which factors may in fact significantly influence the siting of a project.

Morphological, geological, hydrological, topographical and terrain conditions, as well as technical and manpower availability, are among the most important factors affecting reservoir-site selection. The ideal morphology for a reservoir site would be a narrow valley-gorge for the dam, with a flat, broad valley for the reservoir on its upstream side. In practice, however, sites of such nature are rare; but since design work is a steady quest for the optimum compromise among conflicting factors, the morphological aspects would play a very important role in the overall reservoir design exercise. Geology is also of decisive importance in reservoir design; in dam site selection, it is advantageous and also profitable to look for the highest possible elevation of solid rock, thus reducing excavation work. Additional geological considerations require that the site not be on formations that leak excessively, and that there not be any risk of large landslides, rockslides, or rockfalls into the reservoir. The role of hydrology in the reservoir design exercise pertains to the determination of the design flood discharge for spillway and diversion structures, as well as determination of capacity (to include seepage/infiltration, wind-caused wave effects, evaporation and sedimentation issues) and guaranteed water yields.

Ultimately, to obtain economical storage capacity, a reservoir site should be wide in comparison to the dam site and should be on a stream having a low or gentle gradient in order to obtain a long reservoir in proportion to the height of the dam. In general, the location and physical description of the project area(s), together with alternative layouts should be carefully evaluated. There is the need to conduct a major program of subsurface investigation drilling activities, soil and rock sampling and analyses, and geophysical surveys at the site or area in order to generate adequate geological data for the conceptual design studies. Other considerations may consist of population surveys/studies of the project area, community resettlement programs, agricultural development plans, etc.

3.1.2. THE CASE FOR MULTIPURPOSE RESERVOIRS

The design of multipurpose reservoirs present complications when quantities of water released at any time may or may not be made to serve more than one purpose (see, illustrative example in Figure 3.1). Indeed, multipurpose reservoirs are generally called upon to satisfy diverging requirements in accordance with the variety of objectives served.

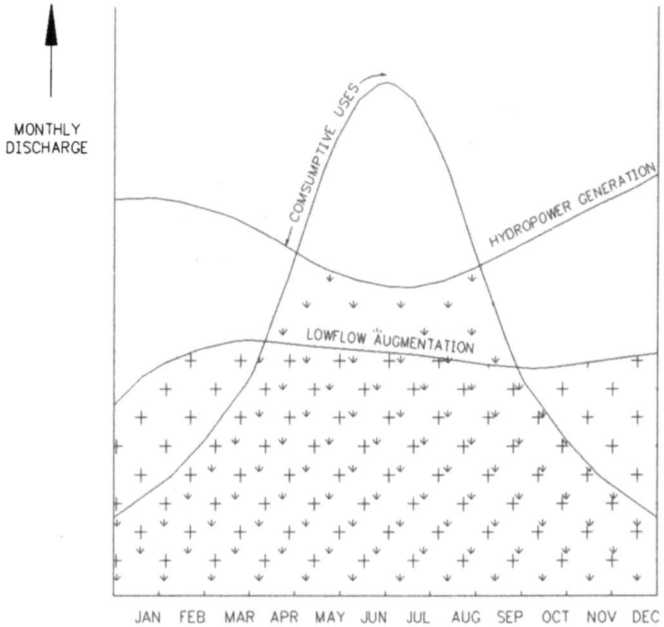

Figure 3.1. Scheduling water releases for different purposes (adapted from Buras, 1972 and Hall, 1964)

Planning for a single purpose in a single reservoir analysis is the simplest task because there tends to be no conflicting demands to be evaluated. Multipurpose reservoir analysis, on the other hand, generally requires detailed sequential analysis to avoid conflicts among purposes. The selection of time interval for analysis depends on the variability or randomness of the supply and demand. Priorities among purposes must be established during the analysis; flood control usually has the highest priority.

In fact, the conservation storage in a reservoir may provide some flood protection – albeit its primary purpose is to provide water whenever demand are in excess of supply, and to provide head for hydroelectric energy. The upper and lower limits of conservation storage may vary seasonally. If several conservation purposes exist, then priorities will have to be established to ensure that the highest priority demands are met. The conservation storage can be subdivided by allocating a portion of the total storage to a buffer zone; when water in storage gets down to the buffer zone, only the highest priorities would be supplied, or supplies for all purposes will be reduced.

The simulation strategy for a multipurpose reservoir can be performed using period-of-record flow data or by using isolated critical flow periods. For flood control analysis, the time interval is generally relatively small and local events are usually used. For most conservation purposes, a monthly time interval is sufficient to evaluate the reservoir operation in planning studies. In general, shorter sequenced critical periods could be used to develop plans or screen alternatives; then, with just a few alternatives, a longer sequence should be used to evaluate performance over a range of hydrologic conditions.

Besides the unquestionable advantages, one major problem with multipurpose storages is the need for coordinating a variety – sometimes conflicting – interests; for instance the conflicting demands of flood control and other uses for storage space tends to create substantial difficulties. In most practical situations, these goals need to be optimized. Accordingly, in planning for multipurpose reservoirs, the site and location of the dam, the capacity of the reservoir, and the allocation of costs must be so-determined as to reconcile – as far as possible – the conflicting interests.

3.2. Multireservoir System Layout and Analyses

A multiple reservoir decision arises when there are two or more reservoirs operating for a common purpose. Although reservoir releases in a multireservoir system can be established on an individual reservoir basis, increased benefits can generally be realized by using a systems approach to determine the releases.

All reservoir systems can be characterized as combinations of tandem (in series) and parallel reservoirs. For *tandem reservoirs*, the release to meet downstream demands is made from the downstream reservoir while releases are made from the upper reservoir to meet local needs and to keep all the reservoirs at some predetermined balance; one exception to this would be when the upper reservoir is a hydropower project. Ideally, the upper hydropower project should only release for power generation while the lower reservoir re-regulates the flow for downstream requirements. The analysis of reservoirs in series is limited by problems of serial correlation between reservoir inflows that are analogous to the serial correlation of inflow between periods. For *parallel reservoirs* operating for a common downstream point, any combination of releases that provides the required total release would be satisfactory to the downstream user; the question to be answered is how much to release from each. In fact, when parallel reservoirs serve the same purposes, the apportionment of required releases among the reservoirs becomes a major decision.

For a system of reservoirs within a river basin, reservoirs may be considered as either acting independently or acting jointly. A reservoir is said to be acting *independently* when the damages and/or shortages to water excesses (or floods), and water shortages (or droughts) can be identified with releases made from that reservoir.

An independent reservoir may be said to be a storage reservoir acting in isolation, i.e., a storage facility operated independently of any other reservoir(s). For instance, referring to Figure 3.2, reservoirs 1 and 4 may be identified as acting independently because all damages due to floods or water shortages between reservoirs 1 and 2 are ascribed to reservoir 1 and all losses/damages immediately downstream of reservoir 4 are attributed to reservoir 4. A set of reservoirs is said to be acting *jointly* when the benefits and damages from them can be identified only with the joint operation of the set as a whole and cannot necessarily be attributed to the individual reservoirs in the set as such. Thus, a dependent reservoir is a cooperating one, and whose design and operation are dependent upon other reservoirs. Referring to Figure 3.2 again, reservoirs 2 and 3 may be classified as acting jointly since the damages/losses below the confluence of the two streams and the damages upstream of reservoir 3 cannot be identified separately with reservoir 2 or 3. The cooperating effect or dependency of reservoirs may arise if, for instance, reservoir inflow depends partly or wholly on the regulated outflow of upstream reservoirs; reservoir outflow is imposed by joint operation with downstream reservoirs; or reservoir outflow is affected by water released from reservoirs in adjacent river basins.

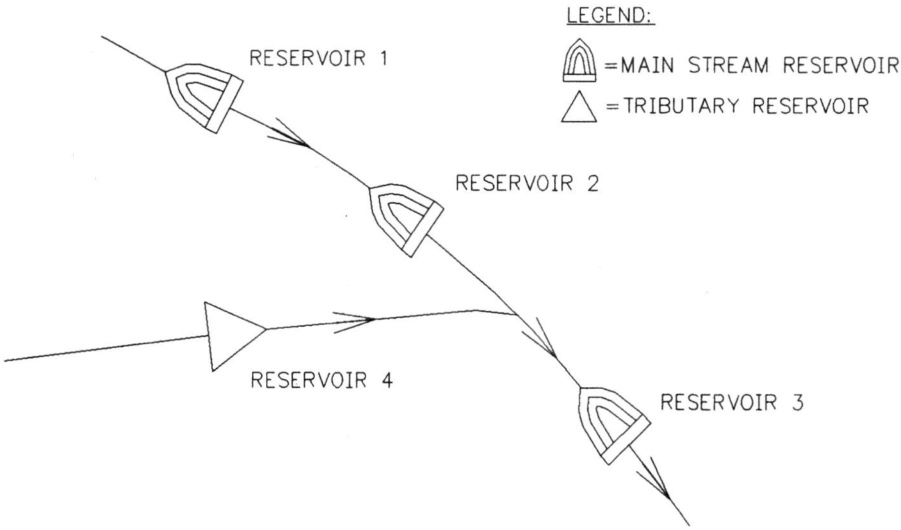

Figure 3.2. A schematic of reservoirs acting independently and jointly for a hypothetical river basin

Figures 3.3 show a hypothetical but representative and typical river basin system – in which reservoirs may be placed in parallel and in series, and in which reservoirs may be acting jointly or independently.

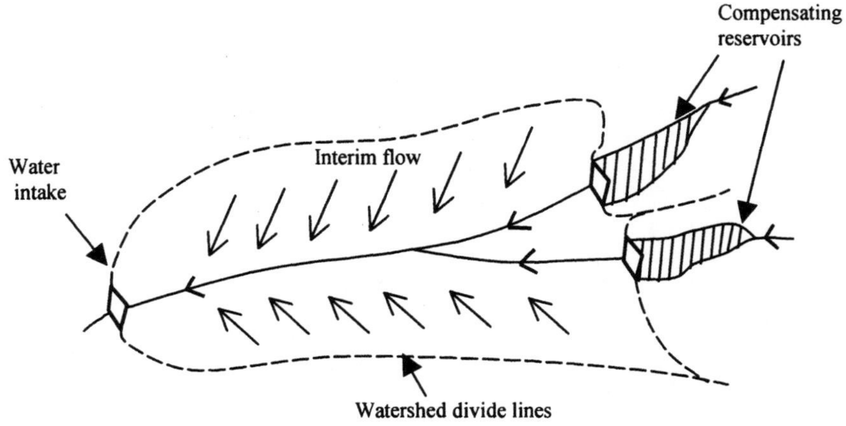

Figure 3.3. Illustrative sketch for typical network of compensating reservoirs

3.3. Hydrological Basis for the Determination of Reservoir Storage Capacity

The primary purpose of storage reservoirs is to provide a means of surface water regulation, both with respect to time and amount; the reservoir storage capacity, together with its operating policy, provides an indication of the extent to which streamflow can be stored for later release or other uses.

Although physically inseparable in some cases, distinction is often made between the various reservoir storage components. In general, reservoir storage capacity can be divided among three major uses (Figure 3.4), namely:

1. The *active storage*, used for streamflow regulation and for water supply;
2. The *dead storage*, required for sediment collection, recreational development, and/or hydropower production; and
3. The *flood storage capacity*, reserved to reduce potential downstream flood damages.

Often these components of reservoir capacity can be modeled separately, and then the total reservoir capacity, K, calculated as the sum of the dead storage capacity (K_d), the active storage capacity (K_a), and the flood storage capacity (K_f). However, since the required active storage capacity varies throughout the year, and since flood storage may not be required in all seasons, a less conservative estimate of the total capacity (K) will be the sum of dead storage and the maximum required active storage and flood storage in the flood season *or* the maximum required active storage in non-flood seasons, whichever is greater. Thus, with an initial storage volume of S(t) at the beginning of period *t*, the following relationships hold true (Loucks *et al.*, 1981):

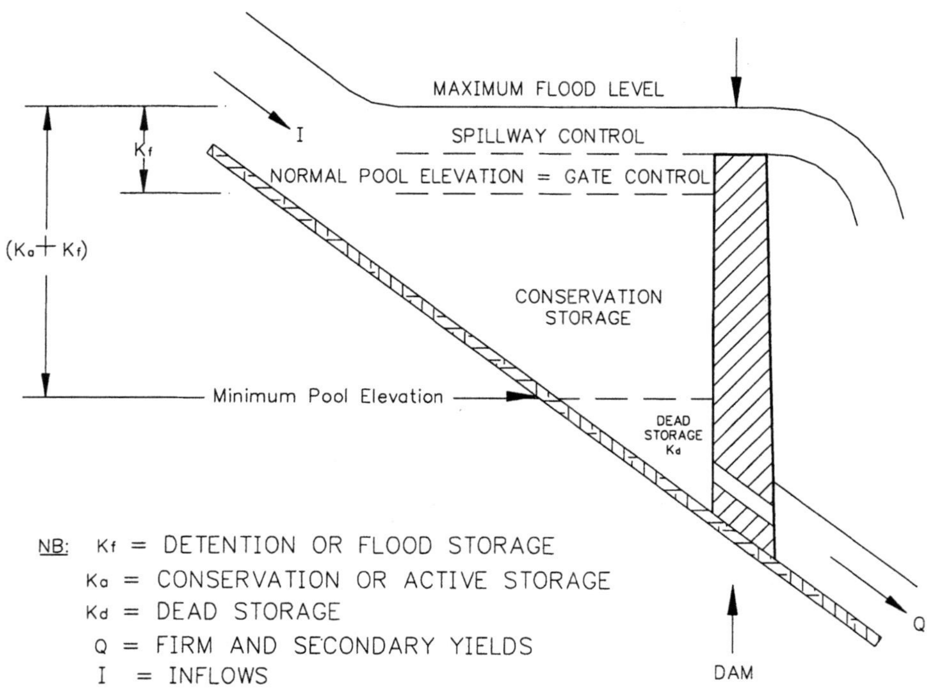

Figure 3.4. **Reservoir storage zones in multipurpose reservoirs**

$$K \geq K_d + S(t) + K_f \qquad \forall t \text{ in Flood Season} \tag{3.1}$$

$$K \geq K_d + S(t) \qquad \forall t \text{ in Non-Flood Season} \tag{3.2}$$

This definition of total capacity permits the trade-off between active and flood storage capacity when and if there can be a trade-off in different seasons (Loucks *et al.*, 1981). Indeed, since reservoirs are designed to reduce flood damages and also to meet low-flow augmentation demands, one must simultaneously take into consideration the drought period inflows and the flood year inflows in any attempts to find optimal capacities for the reservoirs; this helps to side-step the problem of over-design which could result by computing the total capacity via the conservative approach which is based on the 'blind totals.' Although, strictly speaking, the flood attenuating effect of multipurpose reservoirs cannot be enhanced unless the reliability of supply for other purposes is sacrificed, a balance can nevertheless be drawn among all the uses in an optimal design scheduling. It is also noteworthy that, the storage capacity reserved for sedimentation may be utilized for a considerable period after the construction of the dam; thus, in some cases, the water in the dead storage below the sedimentation level may also be utilized in abnormally dry seasons.

The analysis of all possible alternative configurations in the hydrological dimensioning of multiple reservoir systems is indeed a formidable task. In general, any site-/spatial location-based advantage of one reservoir over the others should be very carefully evaluated, and 'priority' given to this factor for utilization in the design process. Ultimately, the project should be economically justifiable, and should be formulated so as to maximize net economic benefits whiles providing a reasonable guarantee of dependable water demands from the reservoirs, and also concurrently satisfying the necessary flood protection, etc.

3.3.1. A MODEL OF CAPACITY ALLOCATION AND SURVEY OF WATER DEMANDS IN MULTIPURPOSE RESERVOIRS

The management of an integrated or multipurpose water control system presents a different kind of problem to that of a single purpose one. A typical allocation of storage in a multipurpose reservoir includes flood control storage, conservation storage, hydroelectric power storage, sediment storage, and a buffer storage. In fact, water resources are controlled and regulated to achieve a wide variety of objectives – and a single purpose reservoir may be considered as a special case of a multipurpose one.

A model diagram of capacity allocation in a typical/representative multipurpose reservoir is provided in Figure 3.5; Figure 3.5(a) shows the case when flood control requirements are less severe, whereas Figure 3.5(b) depicts situations where the requirements for flood control are rather critical. Although sometimes physically inseparable, distinction is usually made between the different components of reservoir storage. In each case, the determination of storage requirements is based on somewhat different hydrologic data. A brief general survey of the demands of some different water uses and capacity allocations, together with a discussion of the hydrological basis for the computation/estimation of storage requirements is presented below for the rather pertinent and commonly designed-for uses.

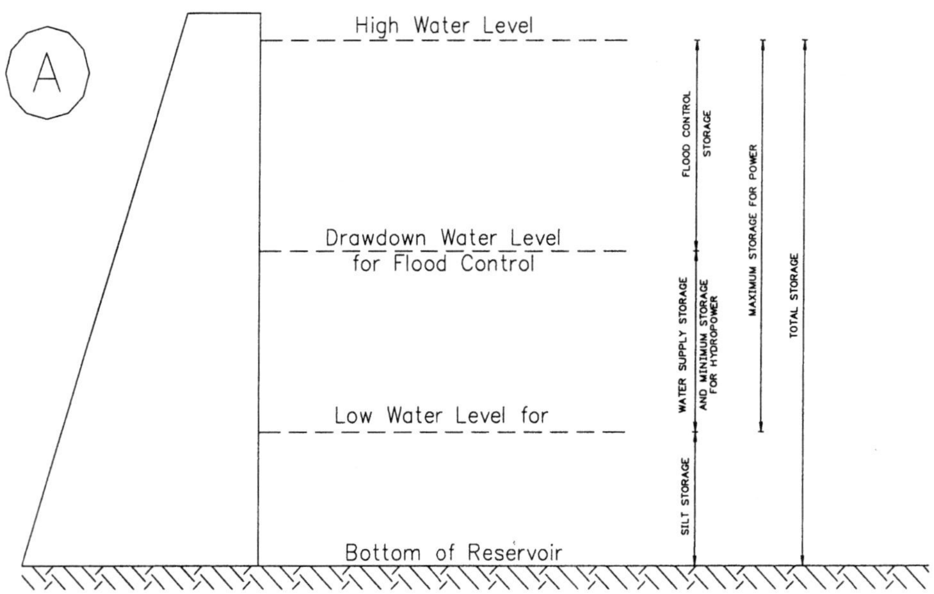

Figure 3.5(a). **Allocation of reservoir storage capacity (adapted from ECAFE, 1962)**

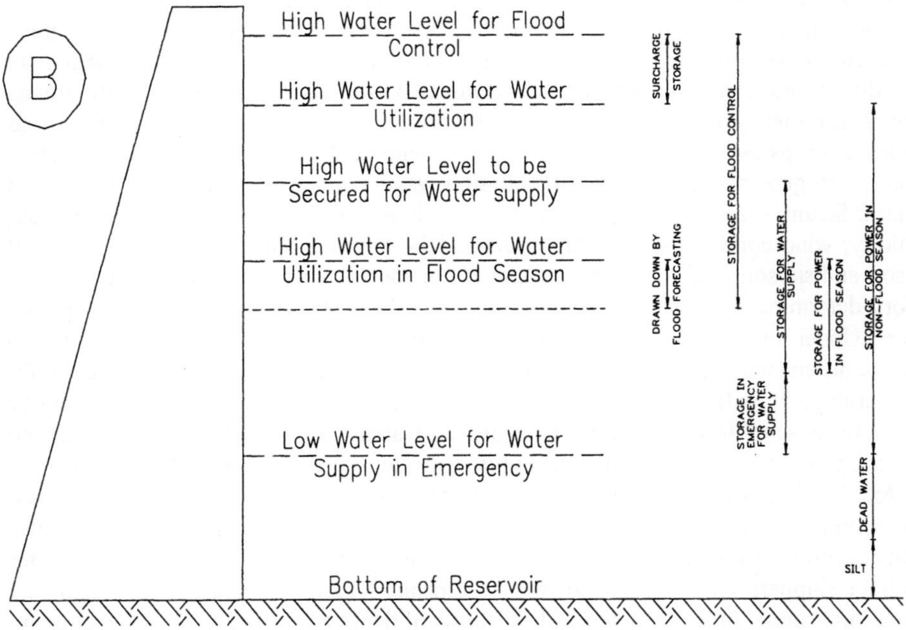

Figure 3.5(b). **Allocation of reservoir storage capacity (adapted from ECAFE, 1962)**

3.3.2. ESTIMATING THE ACTIVE STORAGE NECESSARY FOR FLOW REGULATION AND WATER SUPPLY

The determination of the reservoir storage requirements for all conservation purposes other than hydropower is considered under this topic. Several authors (e.g., UNESCO, 1982; HEC, 1977, 1975; Sokolov and Chapman, 1974; Davis and Sorensen, 1969) elaborate specific steps to be considered in estimating reservoir storage requirements for water supply and irrigation.

In general, requirements for public water usage and industrial demands can be estimated based on population forecasts and assessments, as well as industrial surveys. Future requirements due to growing population and anticipated industrial expansion should also be taken into consideration at the initial planning stages. Additionally, it must be recognized that an improved standard of living of a population usually means a further increase in water requirements for domestic water supply.

The irrigation water demand is almost always periodic in nature. Moreover, there are considerable differences in water quantities required during the irrigation periods of individual years, since the demand is dependent on the distribution of rainfall, among other hydro-meteorological factors. Furthermore, the growing seasons of various species of crops are different – and so are the beginning and duration of the irrigation seasons. In general, the quantity of irrigation water required depends on a number of climatic factors – namely, the volume and distribution in time of rainfall, temperature, humidity, wind conditions and related factors; the physical and chemical properties of the soil and subsoil; and the water demand of the specific crops being cultivated. These factors determine not only the volume, but also the distribution in time of irrigation water. Under similar climatic and soil/pedological conditions, the distribution of crop water requirements depends only on the kind of crops grown – and even that may differ considerably. In extensive irrigation schemes and systems, demand fluctuations can be curbed by a suitable combination of crops and areas of cultivation, under a mixed cropping program, leading to an almost uniform water demand vis-à-vis the supply.

Another important component of conservation storage relates to demands for low flow augmentation for navigational and recreational activities, as well as for water quality control. A navigation development must offer long-range dependability in order to attract shippers and promote growth of terminals along the waterway. Also, the navigation channel must be dependable in terms of depth and of the range of fluctuation of water levels, and should also be reasonably free of wide variations in streamflow and the corresponding velocity changes. The objective of limiting the variation in velocities and water levels in the reservoirs is attained by careful routing of discharges through the reservoir system, and by imposing constraints on the rate of change of discharge from the dams. In addition, recreational use, pollution abatement, preservation of scenic and wilderness areas, etc. helps to maintain or improve health and welfare of populations. Indeed, pollution reduces the utility of water for municipal, agricultural and industrial uses and seriously diminishes the recreational and aesthetic value of lakes and reservoirs; hence water quality control has become a very important objective of many water projects.

Some of the commonly used methods for estimating active storage requirements include the Mass Diagram analyses (Fair *et al.*, 1966; Rippl, 1883), the Sequent-Peak procedure (Fiering, 1967; Thomas and Burden, 1963), Simulation and Optimization techniques and analyses (Wurbs, 1996; Loucks *et al.*, 1981; Hall and Dracup, 1975); some detailed discussion of these methods are presented in Chapters 7 through 10.

3.3.3. HYDROELECTRIC POWER POTENTIAL OF STORAGE RESERVOIRS

Reservoirs may be created for the provision of power for economic development, improved living standards, and an overall enhanced quality of life. The major hydrologic difference between the routing studies for hydropower and other project purposes is that, the quantity of water required in a given time period is a function of two independent variables – power demand and effective head. Furthermore, the quantity of water required is a direct function of the power demand, but an inverse function of the head.

In general, the production of hydroelectric energy during any period at any particular reservoir site is dependent on a number of factors – including the installed plant capacity; flow through the turbines; average productive storage head; number of hours in the period; plant factor; and a constant for converting the product of flow, head, and plant efficiency to kilowatt-hours of electrical energy. The energy produced in period t, E(t) [in kilowatt-hours, Kwh], is proportional to the product of the plant efficiency, η, the productive storage head, H(t), and the flow q(t) through the turbines, viz.:

$$E(t) \propto [\{\eta\}\bullet\{H(t)\}\bullet\{q(t)\}] \tag{3.3}$$

For the total realization from a hydropower project, leakage through the dam, as well as head losses through the penstock, turbines, and draft tubes must be accounted for. The tailwater elevation is also to be given serious consideration in power studies – since a 1-meter loss in head may represent a major economic loss (UNESCO, 1982). Also, the large fluctuations of releases required to meet peak load demands can cause severe bank erosion – and this may require studies to determine storage requirements of downstream re-regulation dams, and the resulting effects of tailwater elevations.

3.3.4. STORAGE-SPACE FOR FLOOD MITIGATION: THE RESERVOIR FLOOD STORAGE CAPACITY DESIGN

The storage space reserved for the accommodation of floods generally cannot be utilized for other purposes. This reservoir volume must therefore be reserved to retain a flood wave, whenever it may occur. As an exception to this general rule, a two-fold utilization may be considered, in periods only where – under the prevailing climatic conditions – no flood, or more precisely no major floods are likely to occur; thus, two-fold utilization is necessarily periodical and partial, i.e., it can extend only to a part of this space.

The objective of all flood control operations is to regulate damaging floods to minimize potential damages. In compliance with the multipurpose reservoir operating rules, storage space in the reservoirs is reserved – and during flood runoff periods, this is used to reduce downstream flood crests. During periods of substantial flood flows, flood control requirements always have top priority over *all* other objectives of system operation. Although the primary objective of flood control operation is the reduction of flood crests to non-damaging levels, a secondary objective of providing timely flood warnings is also important. When the flood cannot be regulated to prevent all damages, timely flood warning can result in substantial lessening of the flood damages, and may indeed prevent loss of life.

To determine the optimal capacity of a reservoir from the point of view of flood control, the inflow data for the worst flood year, selected from the historical inflow data of a specified duration, would typically be utilized in the ensuing analysis. (It is noteworthy that, from the point of view of low-flow augmentation, the inflow data will be based on the severest drought period on record.) In the design of flood control reservoirs, it is usually assumed that the period between floods is long enough for the reservoir to be emptied before the arrival of the next flood. However, there are climatic conditions – mainly in areas affected by tropical cyclones – where two major floods may occur in succession. Under these conditions the period between the floods cannot be ignored and has to be considered as a design variable. Typically, the year is divided into a number of equal time intervals; the number of subdivisions should be such that salient features of the inflows, such as peak flows and duration of a flood are reflected in the design with a reasonable degree of accuracy. If the number of subdivisions is not large enough, a certain amount of information on the inflow rates may be lost since, within every subinterval, the inflow is averaged out.

Ultimately, the determination of flood control storage for a reservoir project is based on hydrologic analyses that are governed by a set of project formulation criteria. The guiding formulation principles generally would require that the project should be economically justified, and that the project should *not* result in a significantly increased flood hazards for any flood event that exceeds the design capacity. The fundamental considerations in the design exercise are elaborated below.

Analysis and Synthesis of Flood Flows: Floodflow Computations

The probability or likelihood of a flood peak of given magnitude is often described by its *return period* (also called the *recurrence interval*). The *return period*, T, of a flood is the expected number of years before the occurrence of a flood of equal or greater magnitude; the probability that a T-year flood will be exceeded in any given year is 1/T.

If Q_p is the random annual peak flood flow, and Q_{pt} is a particular peak flood flow having a return period of *T* years, then the probability of Q_p equaling or exceeding Q_{pt} is given by:

$$\Pr\{Q_p \geq Q_{pt}\} = \frac{1}{T} \tag{3.4}$$

This is the *exceedance probability*, and its distribution function for continuous distributions is defined as follows:

$$\Pr\{Q_p \geq Q_{pt}\} = 1 - F_{Qp}(Q_{pt}) \tag{3.5}$$

or

$$F_{Qp}(Q_{pt}) = 1 - \frac{1}{T} \tag{3.6}$$

The expected annual flood damage at a potential flood damage site can be estimated from knowledge of the exceedance probability distribution of peak flood flows and the resulting damage. The peak flow, Q_{pt}, at any potential damage site resulting from a flood of return period, *T,* will be a function of the upstream reservoir flood storage capacity, K_f, and the reservoir operating policy, viz.:

$$Q_{pt} = f_T(K_f) \tag{3.7}$$

Assuming a known operating policy for flood flow releases, this function, $f_T(\ldots)$, can be defined by routing a series of floods through the upstream reservoir that has various flood storage capacities, K_f, and a known operating policy to the downstream potential damage site. This will require a flood routing simulation model (see, e.g., Viessmann *et al.*, 1977). The flood hydrograph or peak discharge that is ultimately adopted as the basis for design of the structure is called the *design flood*; the selection of the appropriate design flood does indeed require a hydro-economics analysis of the problem.

Flood Control by Multipurpose Reservoir Systems

For the preliminary estimation of flood storage capacities in multi-reservoir systems, and that also involves several potential damage sites, a single design flood can be selected. Using this design flood, the system may be simulated to determine the effect that any combination of flood storage capacities has on reducing the flood peak at some downstream potential damage site(s). The difficulty in this exercise is in obtaining a simple expression for this effect as a function of all the upstream reservoir capacities – hence the need to introduce the concept of an *equivalent flood storage capacity* at a fictitious or actual reservoir site just upstream of the potential damage site. The *equivalent flood storage capacity*, $K_{i,f,}$ at reservoir site *i* will be that capacity which reduces the flood peak at the downstream damage site by the same amount as the actual flood storage capacities, $K_{i,f}$, in the upstream reservoirs at sites $i = 1, \ldots\ldots\ldots, n$ {*n* being an integer indicating the reservoir sites}. This principle of 'equivalent flood storage capacities' is demonstrated in a comprehensive manner in Loucks *et al.* (1981). The flood storage capacities, $K_{i,f,}$ may indeed represent conservative estimates of the capacities needed to reduce flood peaks downstream if active storage capacity or dead storage capacity is included in the total reservoir capacity. This is especially so because existing lakes will tend to increase channel widths and some flood storage may be provided by unused active reservoir storage capacity at the time of a flood – both of which conditions result in reduced flood peaks downstream, even without additional flood storage capacity at the reservoir sites.

In general, for flood protection in a multipurpose reservoir, a certain volume of the storage reservoir is kept empty – ready to receive water during the peak of the flood. The stored water is discharged after the flood peak has passed. The reservoir receives all the water in excess of a design flow rate, Q_d – i.e., the peak of the hydrograph is 'skinned off' and is delayed by storage. Thus the flow rate downstream is limited to a predetermined maximum value. The stored volume is discharged when the flow rate downstream falls below Q_d. This reduces the peak discharge rate of the downstream hydrograph but increases the base-length of the hydrograph.

Controlled vs. Uncontrolled Storages

The reservoir storage may be controlled or uncontrolled. The storage above a spillway crest (i.e., without gates) is an uncontrolled storage – and its effect on the outflow is determined in accordance with the principles of reservoir routing; it is noteworthy that, even this type of reservoir has a control facility in the form of gated bottom outlets –

i.e., sluice gates. For a fully controlled storage, the outlet must be gated – and the discharge capacity at zero storage must be equal to or greater than the maximum possible rate of inflow.

A fully controlled storage is the most effective way of reducing flood peaks, since only the volume above the design discharge needs to be stored (area [1] in Figure 3.6); for an uncontrolled storage, volumes corresponding to areas [1] and [2] have to be stored. Furthermore, the controlled storage is emptied in time t_e after the inflow has dropped below Q_d. The shape of the falling part of the Q_c hydrograph depends on the gate and operating characteristics. The emptying time for the uncontrolled storage is measured by the interval between the point of intersection of the I and Q_u hydrographs, and the point where the recession limbs coincide.

Associated with the higher efficiency of controlled storages is some reduction in the overall safety, since gates may fail to open. Hence, an ungated emergency spillway is usually included in controlled storage schemes. Klemes (1973) elaborates a graphical design procedure for the estimation of the controlled storage volume for a given return period and discharge rate, Q_d.

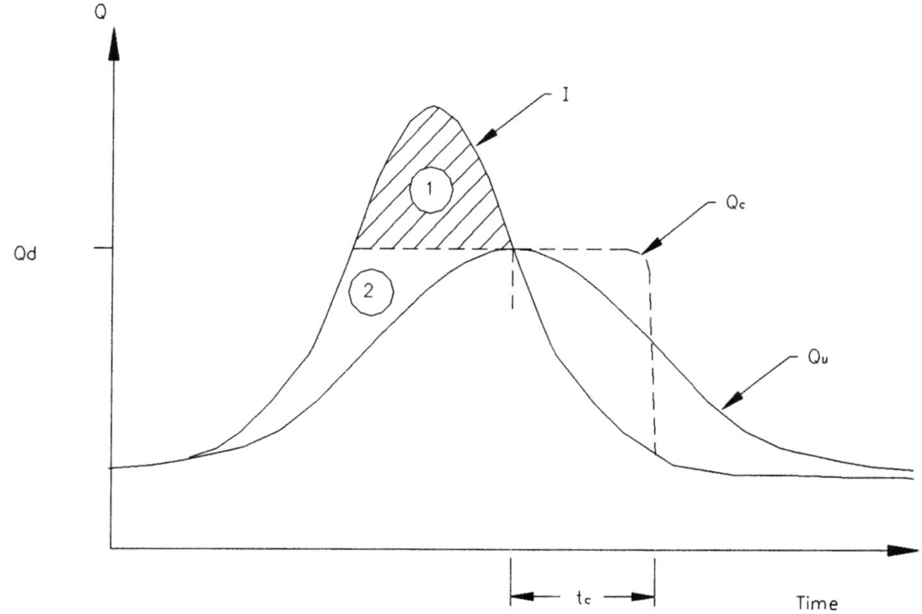

Figure 3.6. Storage for flood control
(hydrographs Q_c, Q_u and I represent the controlled discharge, uncontrolled discharge and inflow, respectively)

3.3.5. SILTATION OF RESERVOIRS AND SEDIMENT RESERVE STORAGE

A major problem associated with the design and operation of reservoirs is siltation. In fact, unless carefully constructed and maintained, the ultimate destiny of all reservoirs is deterioration and failure or filling by sedimentation. If the sediment inflow is large compared with the reservoir capacity, the useful life of the reservoir may be very short. Reservoir planning must therefore, of necessity, include consideration of the probable rate of sedimentation – in order to determine whether the useful life of the proposed reservoir will be sufficient to warrant its construction.

Erosion and sedimentation processes are indeed parts of the geological evolution of landscapes, of which rivers and river processes play a significant role. Also, major engineering works cause a redistribution of water and sediment that affects channel processes. Dams and reservoirs interrupt and alter the natural flow of water and sediment through the drainage system. The reservoir is a sink producing hydraulic conditions that reduce the velocity of flow, often to near zero, within the reservoir. The change in velocity is usually sufficiently rapid as the flow enters the reservoir. The bed load sediment particles settle out first and the suspended sediment particles will travel farthest into the reservoir before settling out (V. Nagy, 1990). Since hydraulic structures on river channels disturb natural processes of riverbed adjustment, any hydraulic project requires the consideration of both the regime of sediment transport and the hydraulic structure's effect on sediment motion.

Ultimately, a reservoir collects sediment, and the eroding capacity of the downstream flow tends to increase compared with that under unchanged conditions. This causes more intensive local erosion adjacent to the dam as well as an intensification of channel processes further downstream (USBR, 1977). These downstream changes may occur over long distances (>10 km); aggradation is intensified upstream in the reservoir backwater, thus producing a rise in water levels.

Problems of sediment transport enter the reservoir design in several ways. The designer has to consider the response of the river to the reservoir upstream and downstream from it, the siltation of the reservoir itself, and the handling of the river during the construction of the reservoir. It is noteworthy that, prediction of the rate of sediment supply to the reservoir and the amount retained in the reservoir is still a highly empirical art.

The Nature and Extent of Reservoir Sedimentation

Soil erosion and sedimentation is a universal problem in the design and management of storage reservoirs. In any case, sediment yield attributable to Asia is estimated to be more than twice the world average. For instance, the maximum sediment load can be as high as 1600 kg/m^3 in the Yellow River and its tributaries and the Yongding River in China – and this has resulted in serious sediment deposition problems in the reservoirs constructed on these rivers.

In fact, in some of the worst cases of reservoir sedimentation on record, more then 50 percent of the storage capacity had been lost after few years of operation. The average annual rate of loss of the storage capacity (i.e., ratio of average annual deposition volume to total storage volume) has been estimated at 2.3 percent (Chun Hong, 1995). Stall and Lee (1980) have reported that in some reservoirs in Illionois (USA), the rate of capacity loss is about 4.8%. Also, the 76 m high Warsak Dam on the Kobul River in Pakistan lost 18% of its storage volume in the very first year of operation. Furthermore, Pleskov (1985) provides detailed characteristics of the sedimentation of the six reservoirs constructed on the Don River in Russia (Table 3.1).

As a further example, the case of the Tiszalŏk barrage in Hungary with a storage space of 10 x 10^6 m^3 is worth mentioning here. In this case, the measured suspended sediment transport of the Tisza River fluctuates between 1 and 9 million tons annually. The discharge characteristics of the river are: Q_{min} = 56 m^3/s, Q_{mean} = 530 m^3/s, and

Q_{max} = 3620 m^3/s. The flow velocity is around 15 - 25 cm/s in the main channel, and between 0.5 - 10 cm/s in the flood plain. During the first two years of operation, an aggradation of 7 m was observed along certain reaches, and in the first 7-year period, 4.5 x 10^6 m^3 deposition was registered in the reservoir; indeed, it has also been reported that the actual deposition during this period was 5.7 x 10^6 m^3, except that 1.2 x 10^6 of the total volume became stirred up and transported away by flood waves (Rákoczi, 1994).

Table 3.1. Characteristics of the Don River reservoirs

Reservoir characteristic	Reservoir identification number					
	1	*2*	*3*	*4*	*5*	*6*
Area of the watershed (km^2)	430	403	130.5	271	540	1113
Average runoff, mill. m^3	13.6	23.5	8.2	17.5	34.0	45.6
Volume of the reservoir, mill. m^3	25.0	26.8	8.5	16.3	1.3	52.0
Total volume of sedimentation 10^6 m^3	988	687	273	436	1000	4000
Time of operation, year	7	7	7	5	7	11
Average sedimentation 10^6 m^3 / year	141	98	39	87	143	364

Prediction of Sediment Yield
A large variety of erosion-sediment yield models are available in the literature – with the commonly used methods for prediction of sediment yield elaborated below.

- **USLE.** The universal soil loss equation (USLE) method (Smith and Wischmeier, 1957) computes the soil loss at a given site as a product of six major factors, as follows:

$$A = R \cdot K \cdot L \cdot S \cdot C \cdot P \tag{3.8}$$

where, A is the soil loss per unit area, R is the rainfall erosivity factor, K is the soil erodibility factor, L is the field length factor, S is the field slope factor, C is the cropping management factor, and P is the conservation practice factor. This equation has been used to estimate sediment yields for the design of small reservoirs (Haanet *et. al.*, 1982). Simons and Sentürk (1992) indicate that the USLE is the most widely used regression model for predicting soil erosion.

- **MUSLE.** The modified universal soil loss equation (MUSLE) was proposed by Williams (1975) by replacing the factor R in the USLE model with a runoff factor, as follows:

$$A = 11.8 \cdot (V_Q \cdot Q_P)^{0.56} \cdot K \cdot L \cdot S \cdot C \cdot P \tag{3.9}$$

where V_Q is the volume of the runoff (m^3), Q_P is the peak flow rate for the storm (m^3/s), and the other terms have the same meaning as in the USLE method above. This model actually considers both surface erosion and sediment movement in the catchment.

The factors in equations 3.8 and 3.9 (R, V_Q, Q_P), and accordingly the A value, are random variables and consequently the amount of sediment transported by the rivers is also a random variable. Therefore, for the prediction of sediment transport in the rivers, some empirical probability distribution functions should preferably be utilized (Mosaedi, 1998). Indeed, several variations of the above methods, as well as various other methods for the prediction of sediment yield are available in the literature of soil science, hydrology, and water resources.

Reservoir Trap Efficiency

Reservoir sedimentation rates can be estimated by surveys to determine the rate of sediment accumulation in reservoirs that have been in existence for many years. The actual accumulation of sediment in a reservoir depends on the percent of the inflowing sediments that is retained in a reservoir, defined as its *trap efficiency*; this is a function of the ratio of reservoir capacity to total inflows. In fact, the value of the trap efficiency depends on the sediment fall velocity, rate of flow through the reservoir, the geometric features of the reservoir and outlet structures, reservoir operation, as well as on the age of the reservoir; it can also be affected by chemical properties if flocculation occurs. Although trap efficiency of large reservoirs tends to be high, it does not increase linearly – and the useful life of a large reservoir is generally longer than that of a small reservoir, if all other factors remain constant.

In general, the sedimentation process may dictate the lifetime of a reservoir. The lifetime of the reservoir is that time period when the dead space of the reservoir are filled whiles meeting the target demands all the same. It is assumed that, at the beginning of the reservoir operation, the dead space will get filled up first, and then after that, the sedimentation of the serviceable space of the reservoir will follow. Thus, the duration of sedimentation of the dead space is defined by: $t_d = V_d/W$ – where V_d (m^3) is the volume of the dead space and W (m^3/year) is the multiannual average volume of the sediment discharge; and the duration of full sedimentation of the reservoir is $t_f = V_{H,max}/W$ – where $V_{H,max}$ (m^3) is the volume of the reservoir that is associated with the maximum water level. In continental conditions, the typical duration of the full sedimentation is about 20 to 30 years for the local small reservoirs used for drinking and industrial water supply; in the case of larger reservoirs, the expected duration is about 50 years (Pleskov, 1985). If $t_f < 50$ years, then the process of sedimentation could become a serious design parameter; consequently, this factor should be carefully analyzed, because the existence of this type of situation means sedimentation can be very intensive. If t_f is 50 to 200 years, then a far less detailed evaluation will typically be adequate.

The volume of the sedimentation can be evaluated on the basis of measured volume of bed-load and suspended sediment. The distribution of sediment within the reservoir and the retention of sediment depend on sediment size and texture, reservoir inflow and outflow and the size and shape of the reservoir. Estimation procedures for these and related parameters are discussed in the literature by various authors (e.g., Linsley *et al.*, 1982; UNESCO, 1982; Gottschalk *et al.*, 1971; HEC, 1977; Vanoni, 1976; Davis and Sorensen, 1969; Chow 1964).

Reservoir Sediment Storage

The most common procedure for dealing with the sediment problem is to designate a portion of the reservoir capacity as *sediment storage*. This may be viewed as a 'negative' approach that in no way reduces the sediment accumulation problem but merely postpones the date when it becomes serious; nevertheless, this is a reasonable and easy procedure to employ in the storage capacity estimation/design process.

Overall, the fundamental reservoir design question of: 'how large does the reservoir storage need to be, in order to provide for a given demand with an acceptable level of reliability?' calls for need to carefully establish the appropriate relationship between inflow characteristics, reservoir storage capacity, reservoir release, and reliability of operation (Mosaedi, 1998). In this effort, sediment storage and related effects should also be very carefully assessed. In fact, many reservoirs cannot perform as designed because much of their storage volume gets filled with sediment too soon. Bed-load sediment particles may also cause abrasion of turbine blades, tunnels, and gate recess. Furthermore, accumulations of sediments in various parts of the gate structure may affect the operation of these structures. Every effort should therefore be made to minimize these potential shortcomings.

After a site has been selected, the reservoir capacity should be made large enough to create a useful life sufficient to warrant the construction of the dam. Sediment reserve storage is generally estimated separately; this dead storage should be large enough so that it will not be filled before the economic life of the project is over. Indeed, the long-term storage of sediment in a reservoir is stochastic in character and its prediction must – more realistically – therefore be based on the theory of probability. Gottschalk *et al.* (1971) discuss a typical stochastic sediment storage model that is formulated based on Moran's theory (Moran, 1959); the total sediment deposited in the whole period is indicated as an additive process defined on a finite Markov Chain.

It is noteworthy that, the vulnerability of the economic life of a reservoir to sedimentation problems stems from the nonlinear relationship that exists between yield of water and reservoir capacity. Severe reductions in yield of water from a reservoir can result from relatively small changes in reservoir volume – and the reduced storage space could be the result of reservoir siltation.

3.3.6. ADJUSTMENT OF STORAGE ESTIMATES FOR NET EVAPORATION LOSSES

Supplementary evaporation associated with storage reservoirs is the result of the increased water surface following the creation of the reservoir; consequently evaporation from the water body is greater than that from the pre-construction surface. Prior to dam construction, the long-term rainfall-runoff relation for the area that will be flooded by the proposed reservoir can be expressed as follows:

$$R_b = P_b - ET_b \tag{3.10}$$

where R_b = mean annual runoff before reservoir inundation; P_b = mean annual rainfall; and ET_b = mean annual evapotranspiration. After the reservoir is filled, the relation can be expressed as:

$$R_a = P_a - EO_a \tag{3.11}$$

where R_a = mean annual runoff after reservoir is filled; P_a = mean annual rainfall; and EO_a = mean annual evaporation from the water surface. Assuming that the mean annual rainfall before and after construction remains approximately equal, i.e., $P_b = P_a$, then:

$$\Delta E = R_b - R_a = EO_a - ET_b \tag{3.12}$$

where ΔE = mean annual net evaporation loss, and noting that usually $\Delta E \geq 0$.

Anticipated evaporation is a particularly decisive element in design of reservoirs to be constructed in arid regions. Rates of evaporation vary, depending on meteorological factors and the nature of the evaporating surface. The relative effect of pertinent meteorological factors is difficult to evaluate at best – recognizing that the rate of evaporation is influenced by solar radiation, air temperature, vapor pressure, wind, and minimally by atmospheric pressure. Since solar radiation is an important factor, evaporation also varies with latitude, season, time of day, and sky condition. Though salinity effects can be neglected in the estimation of reservoir evaporation, any foreign material that tends to seal the water surface or change its vapor pressure or albedo will affect the evaporation. Possible methodologies to employ in evaporation estimations needed in reservoir design include water-budget determinations of reservoir evaporation; energy-budget determinations of reservoir evaporation; aerodynamic determination of reservoir evaporation; combination methods of estimating reservoir evaporation; and estimation of reservoir evaporation from pan evaporation and related meteorological data (Linsley *et al.*, 1982).

Open surface water evaporation can be estimated by using one of the available theoretical and empirical procedures or by applying the annual pan coefficient (p) to tank evaporation data (E_p), as follows (see, e.g., Chow, 1964):

$$EO_a = p \bullet E_p \tag{3.13}$$

Pre-dam evapotranspiration estimates are difficult to determine; one approach is through equation (3.10) above, which yields the following relationship:

$$ET_b = P_b - R_b \tag{3.14}$$

Estimates of mean annual rainfall (P_b) are readily available; mean annual runoff (R_b) can be calculated using data for the catchment, or estimated from regional runoff maps.

The basic requirements for the estimation of evaporation losses based on storage-area relations for a given reservoir are the reservoir storage volume/surface area function – represented by a storage-area relationship (see Figure 3.7 for illustrative sketch); and the average evaporation rate, e_t, for each period. Multiplying the average surface area [m^2] by the loss rate e_t [m] yields the volume [m^3] of evaporation loss in the period. The evaporation loss L_t may then be approximated by (Loucks *et al.*, 1981):

$$L_t = A_a e_t \left(\frac{S_t + S_{t+1}}{2} \right) + A_o e_t \tag{3.15}$$

where S_t, S_{t+1} = storage volumes [m^3] in the reservoir at the beginning and end of period *t* respectively; and e_t = average evaporation rate [m] for period *t*.

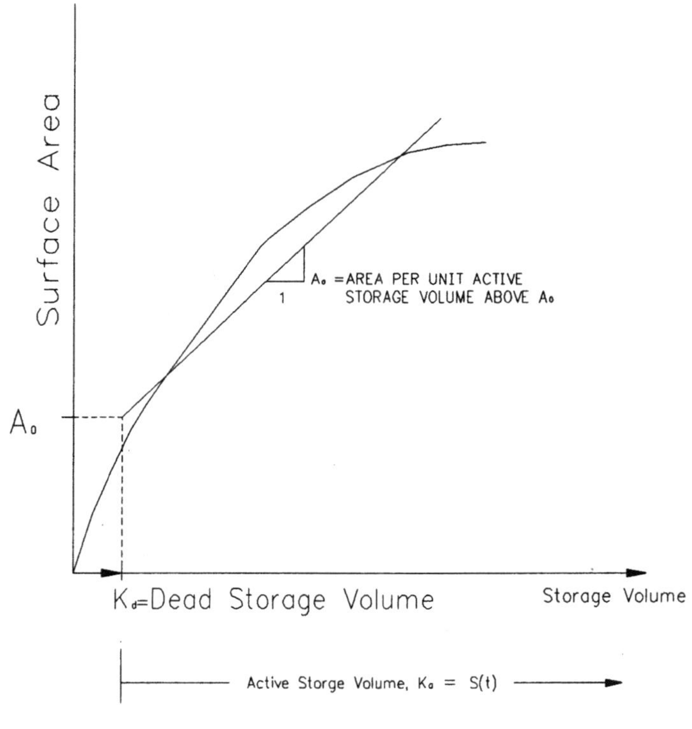

Figure 3.7. A schematic of storage-area relationships and approximation of surface area per unit active storage volume (adapted from Loucks *et al.*, 1981)

In reservoir design, the engineer should be truly concerned with the increased loss over the reservoir site resulting from the construction of the dam, i.e., reservoir evaporation less evapotranspiration under natural conditions. Indeed, it must be recognized that the role of evaporation losses on reservoir design can be rather significant – as illustrated by the shifts in storage requirement curves shown in Figure 3.8. Consequently, the adjustment of storage estimates for net evaporation losses should be carefully evaluated in the reservoir design effort – especially for those projects planned for arid regions. In humid areas, construction of a dam causes only a nominal increase in water loss.

It is noteworthy that evaporation may be reduced by mechanical covers and/or by artificial films if there are no significant wave effects. On the other hand, the presence of certain aquatic plants can bring about tremendous evapotranspiration losses.

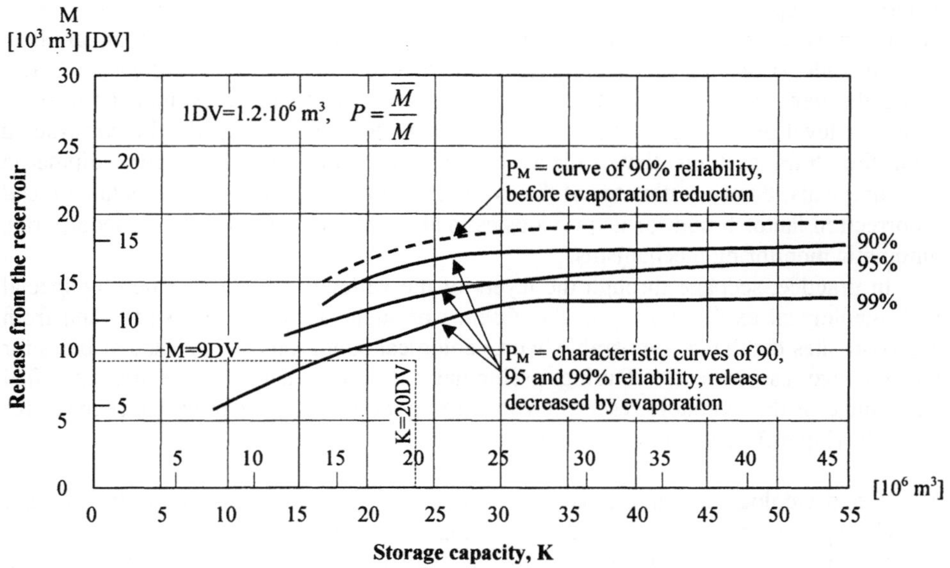

Figure 3.8. Characteristic annual curves with evaporation reductions for the Borjad Reservoir – showing reliabilities of supply

3.3.7. OTHER SECONDARY FACTORS AFFECTING RESERVOIR SIZE-SELECTION

Water losses from a reservoir consist of 'supplementary' evaporation, leakages, and under certain climatic conditions, of water losses due to ice formation. The water losses due to ice formation are connected with the fact that in winter time, when emptying the reservoir, sheets of ice cakes may remain on the shores of the reservoir. Whiles these losses may not be desirable, the benefits derivable from regulation at the reservoir site should offset the hydrologic losses and the costs of reservoir construction and operation – otherwise, the reservoir should not be built at all.

Analysis of Seepage Problems

Further to evaporation losses, the use of reservoirs to temporarily store streamflows often results in a net loss of total streamflow due to seepage. Leakage losses from reservoirs are composed of seepage through the valley bottom and slopes, and through the layers below the hydraulic structures. The loss due to seepage beneath the dam structure is a function of the water head, *h*, and seepage entry surface area, A_s (which in turn is dependent on the storage volume, S). Various more exact and complicated hydraulic models may be used for the calculation of seepage losses; however, it seems that a simplified model based on the Darcy law can be used for a stochastic description of seepage. Making the simplifying assumption that the seepage below the dam follows Darcy's law, then:

$$q = k \bullet i \bullet A_s \tag{3.16}$$

where q = seepage discharge (m^3s^{-1}); k = Darcy's permeability coefficient (ms^{-1}); A_s = surface area at entrance of seepage water (m^2); and $i = \Delta h/L$ = hydraulic gradient (dimensionless), where Δh is the change in reservoir level (m) over distance L (m) along the dam axis. The seepage loss, q, under the dam can be estimated for every reservoir level prevailing at one time or the other. Reservoirs may also be constructed with the drain systems that collect seepage water to the tailrace for volumetric measurements; thence, estimates of the average values of A_s and k can be obtained, and a correction factor found for its previous estimates. Modern systems would incorporate automatic monitoring mechanisms.

In practice, seepage through the layers of the valley can be determined by special hydrogeological explorations. In the absence of such investigation results, data from reservoir sites that have comparable hydrogeological conditions can be used as basis for the seepage assessment. During preliminary design calculations, some specified percentage of the available reservoir volume may be used as the probable seepage for the following different geological situations:

a) 'Good Geological Conditions' – comprising clay loam bed of great strength and close proximity of groundwater on the slopes, use 5 to 10 per cent.
b) 'Average Geological Conditions' – i.e., soil of fair to marked perviousness (in conjunction with the use of anti-seepage measures), use 10 to 20 per cent.
c) 'Difficult/Poor Geological Conditions' – consisting of pervious ground, use 20 to 40 per cent.

In general, where groundwater levels at the dam site are higher than the maximum water levels in the reservoir, no seepage losses from the reservoir would be anticipated. On the other hand, when groundwater levels of the selected dam site are lower than the maximum value of the water level in the reservoir (especially in the case of very coarse sand and coarse gravel or cobbles), seepage losses can be very high – in which case the feasibility of the whole project becomes questionable.

Freeboard Requirements for Wind and Wave Action

Freeboard is the vertical distance between a referenced water surface and the top of the dam structure; it is used to insure the safety of structures against overtopping in the event of the occurrence of the design flood. A dam must normally be constructed with a freeboard above the maximum water depth, H_{max}, associated with the design capacity, K; this would then allow the reservoir/dam to safely pass the floods that would be routed through it, as well as prevent any wind-generated waves from jeopardizing the dam crest.

The purpose of freeboard is to act as a safety measure against overtopping of a structure due to an adverse coincidence of physical forces acting to force the water level to rise above the maximum still water level. Factors that can act independently or in concert to cause this relative rise in water surface above the normal still water design level include sustained high wind, long fetch, ice in spillways, and floods exceeding the spillway design flood.

A typical approach to use in this evaluation of the freeboard requirements would be to determine the crest elevation on the basis of either the highest flood discharge passing the reservoir, or the strongest winds – since the probability of coincidence of these two independent random events is rather small (and can be neglected for all practical purposes). A discussion of some relevant mathematical relationships for estimating the wave height is elaborated elsewhere in the literature (see, e.g., V. Nagy, 1990; Davidan, 1989; Ivanov, 1989; UNESCO, 1982, 1974; USACERC, 1977).

The highest flood flows usually pass reservoirs within a few hours. In Europe, for instance, rarely does the duration of the strongest wind exceeds 4 to 6 hours. The coincidence of the extreme values of these two random variables, flood and wind, can be neglected for all practical purpose – the probability of coincidence of these two independent events being of a small order of magnitude (about 10^{-4}). Thus, a reasonable probabilistic approach to determining the crest elevation is based on one of these two variables, but checked against the other. The distribution functions of reservoir stage and the wave height must be determined in order to establish the crest elevation.

In design practice, the value of the dam crest elevation is usually calculated as follows:

$$H_{\max} = H_{av} + \Delta h + h_w + a \tag{3.17}$$

where H_{av} is the average water depth, m; Δh is the wind-caused water level rise, m; h_w is the height of the wave, m; and a is the safety factor (a ~0,5 m). The maximum value of h_w is usually selected at the probability of occurrence level of between 0.01 and 0.05, depending on the importance of the structure. For the purpose of a preliminary evaluations of the wave height, where the wind speed, v~20 m/s, Figure 3.9 can be used. A more detailed analysis and empirical results are presented elsewhere in the literature (e.g., USACERC, 1977; Davidan, 1989). For shallow reservoirs where the length of the reservoir, L < 100 km, the wind-caused 'super-elevation' of the water surface can be neglected by simply adding a safety value of depth (a), depending on the importance of the reservoir. This value is usually $a \cong 0.5$ m. Subsequently, the dam crest elevation is determined by the height of the wind-generated waves.

Ultimately, the amount of freeboard is partially based on experience – and should be higher when damage potential is greater. Overall, overtopping of the non-overflow sections of the structure should be prevented if such overtopping is likely to cause failure of the embankment.

3.4. Hydrologic Data Requirements and Analyses

The ideal situation in planning water resources development projects is to have a relatively long record (i.e., 30 years or more) of river discharge data at or near the dam site under consideration. Unfortunately, such records are rarely obtainable at or near the desired location, especially in developing regions of the world; under such circumstances, various estimation procedures, extrapolations, and expert professional judgment tends to play a very crucial role in the system design and analysis. In any event, the length of record alone is not a reliable indication of the true recurrence interval of extreme events that fall within a record. As such, a search should generally be made of local historical records and newspaper files for general information on past floods and droughts – in order to obtain a general indication of recurrence intervals, and to complement the existing data.

In general, the pertinent hydrological factors influencing reservoir storage determination include: *inflow variability* – measured by the coefficient of variation; and *serial correlation of inflow* – since the use of an erroneously high correlation would invariably result in an over-design of the storage. The key non-hydrological factors generally would include: *demand levels* – which should be accurately specified in order to avoid either over-design or under-design; and *system life span* – which should be realistic enough to make the project economically viable. Of course, this is by no means the only issues that the systems analyst or designer has to consider in the big picture of storage determination. Indeed, several non-technical issues could become even more important in the ultimate decision about a project. Some relevant and appropriate systems analysis and statistical tools typically used to facilitate the data analyses of both the hydrological and non-hydrological parameters that affect the reservoir design process are discussed below.

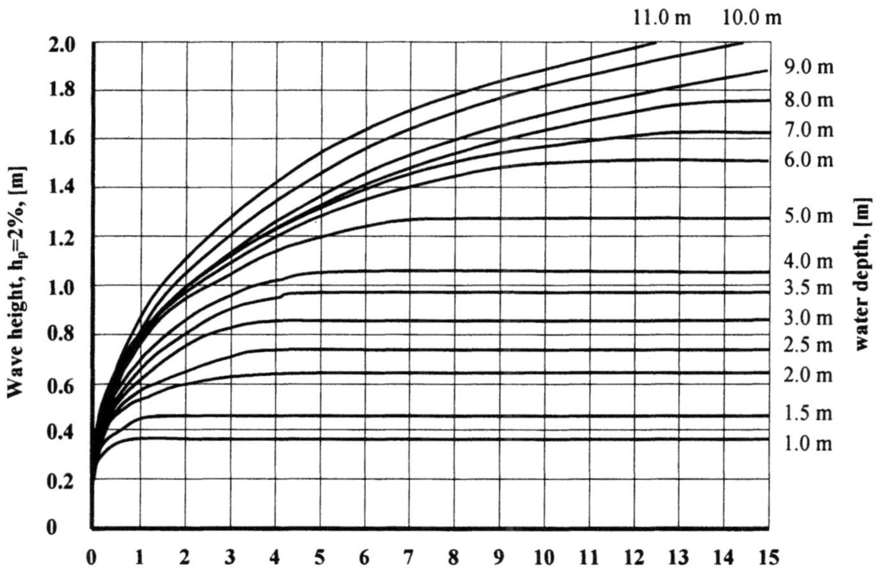

Figure 3.9. Forecasting curves for wave height

3.4.1. SELECTING A DISTRIBUTION FOR USE IN THE HYDROLOGIC DESIGN PROCESS

The probability distributions generally used in hydrologic design efforts are the Normal, Log-Normal, Gamma, and related families – the most widely used, and therefore the apparently most important one being the Normal (also called Gaussian) distribution. The key statistical parameters of these distributions, as necessary for their effectual application in the hydrologic design process, include: the *mean, coefficient of variation, coefficient of skewness,* and *serial correlation coefficient* – which, respectively, represent various measures of central tendency, variability, skewness, and persistence. Detailed discussions of these distributions and their corresponding statistical parameters are presented elsewhere (see, e.g., Kite, 1977; Clarke, 1973; Yevjevich, 1972a; Chow, 1964). In many cases, good estimates of a distribution's parameters are obtained by the 'maximum-likelihood-estimation' (MLE) procedure. Alternatively, the 'method of moments' (MM) gives a quick and simple method for obtaining parameter estimates for many distributions. In general, MLE parameter estimates are to be preferred over those obtained by the MM; computer algorithms (that also supply estimates of the standard error of estimated quantiles) for fitting most distributions used in hydrology by both the MM and MLE procedures are available in the literature, and also on the commercial markets.

During the hydrologic design process, if we assume that any inconsistency and non-homogeneity are identified and removed from the original data series, then estimates of the above-indicated statistical parameters usually will be utilized to fit an appropriate distribution function to the existing time series. In fact, there are several methods by which the sample distribution can be fitted. In any case, despite its scientific basis, the selection of a distribution for use in the hydrologic design process must involve some intuition and common sense. For instance, in trying to select an appropriate distribution, the analyst may start by plotting sample value and considering sample moments. Unfortunately, in many situations, these procedures will not provide an easy choice – because it is quite possible that no one of the identifiable distributions will fit significantly better than the others. The analyst must then consider the purposes of the proposed simulation, and must assess the expected costs of making incorrect decisions. Indeed, there might even be the need to invoke economic considerations further on the selection process; hence, carefully selected benefit functions may be used both to choose the optimal design for a given distribution of natural flows and to choose among possible flow distributions when several seem reasonable and acceptable.

Ultimately, the fitting of a theoretical probability distribution to flow data enables judgments to be made about the probability with which a given runoff amount will be equaled or exceeded – a parameter that has very important implications in the economic planning of projects.

3.4.2. BAYESIAN TECHNIQUES FOR PARAMETER ESTIMATION WITH LIMITED DATA

The application of Bayesian techniques for parameter estimation can potentially improve the available hydrologic data by taking into account not only the information contained in the historical sample, but also all the information coming from other sources – both objective and subjective. Concurrently, project economics can be considered by the use of a loss function that specifies the seriousness of choosing an estimate that is not the true one.

In fact, the classical methods of parameter estimation can only take into account the information contained in the historical sample; they also produce values that are independent of the economic consequences of erroneous estimates. Bayes' theorem seems to provide a framework for addressing these undesirable outcomes. Firstly, an *a priori* distribution on the parameter of interest is provided. This prior distribution encompasses all information that is not in the historical sample, providing an assessment of both the most likely values and the degree of uncertainty of the parameter in question. The prior information enters the estimation procedure via Bayes' theorem, which is expressed by:

$$P(\Theta/Y) \propto P(\Theta)P(Y/\Theta) \tag{3.18}$$

where Y = vector of sample observations; Θ = parameter of interest; P(Θ/Y) = 'posterior' probability density function (pdf) of Θ, given Y; p(Θ) = prior pdf of Θ; and P(Y/Θ) = likelihood function for the parameter Θ. The posterior pdf now replaces the likelihood function as the means of making inferences about the parameter, and as a means for taking into account the second problem of economic consequences of erroneous estimates. These are incorporated by means of a loss function, L(θ,Θ), which specifies the opportunity loss that occurs when Θ, the true value of the parameter, is erroneously estimated as θ. Hence, the Bayesian criterion consists of choosing the value θ that minimizes the expected opportunity losses, viz.:

$$\theta = \min_{\Theta} E_{\Theta}\left[\ell(\theta,\Theta)\right] \tag{3.19}$$

or,

$$\theta = \min_{\Theta} \int_{\Omega} \left[\ell(\theta,\Theta)\right] p(\Theta / Y) d\Theta \tag{3.20}$$

where Ω is the region of the parameter Θ.

Obviously, the Bayes approach depends rather heavily on two factors – the prior distribution, and the loss function. If the prior pdf is to adequately represent all information other than that contained in the sample, then it must be assessed with great care. An important source of information could be to collect past records from other river basins that relate to the parameter of interest. Another source of information could be the analysis of physical characteristics of river basins and their relationship to the parameter of interest; then, by performing regression analyses of these characteristics, a measure of the mean and variance of the parameter can be obtained, and subsequently an appropriate probability distribution can be fitted.

The loss function L(θ,Θ) is the function that specifies the concomitant opportunity loss due to the designer/analyst accepting and designing based on θ – rather than use the real parameter value Θ. These opportunity losses represent the difference between the benefits actually to be obtained from a given project and the greater value that would have been realized had the true parameter value been known. Pratt *et al.* (1965) elaborates a procedure for deriving the loss function.

Indeed, reservoir storage capacity design and analysis is one of the areas that lend itself to potential application of Bayesian techniques for parameter estimation. In the determination of the optimal storage capacity of a reservoir to provide regulated flow for one use or another with specified target demand(s), the necessary action is the construction of the reservoir of capacity K. In this case, the value functions would consist of the net benefits derived from the reservoir operation; these can be computed on the basis of the long-term benefits derived from the operation/regulation, the cost of the reservoir and attendant system components, and the short-term losses which occur due to water shortages. As the reservoir size is increased, the short-term losses decrease at the expense of reservoir costs – and therefore the problem essentially consists of a trade-off between the two factors. Using the assumed flow model with a parameter value Θ_j, and setting the reservoir capacity at K_j, the system net benefits, $NB_t(\Theta_i,K_j)$ can be determined after running a simulation for the appropriate period of time.

3.4.3. DESIGN OF RESERVOIR STORAGE FOR STOCHASTICALLY VARYING WATER DEMAND

To estimate the design capacity of a reservoir, it is necessary to first estimate the demands that will be placed upon the storage at some time(s) in the future – a rather difficult and uncertain task. In most reservoir design procedures, reservoir draft is assumed to be constant or to vary in a predetermined fashion – the variable nature of water demand being at best accounted for by assuming a normal distribution. However, the random variability of hydrometeorological factors, for instance, calls for the need to consider water demand as a random variable – especially when designing a reservoir capacity for supply of irrigation water and for the lowering flood peaks. In fact, future water demand should not be computed deterministically by the use of unique relations as a function of time; rather, the randomness of water requirements should be brought into the overall design picture so that water demands are estimated as random variables. The predicted mean demand as a function of time should be assessed and supplemented by expected variations, measured by the variance or other parameters of dispersion (Linsley and Franzini, 1979; Yevjevich, 1972b).

V. Nagy (1980) presents analytical and graphical procedures to determining the distribution functions of storage capacity for variable water demands – when both inflow and water demand are treated as random and independent variables. Other methodological profiles to the modeling of water demand relations for various uses are outlined in greater detail elsewhere in the literature (e.g., Kindler and Russell, 1984).

3.5. Deterministic vs. Stochastic Methods in the Reservoir Design Problem

During the past few decades, there has been several important developments in analytical and statistical methods used in the design and operation of multipurpose reservoir systems – albeit the basic classical methods for the dimensioning of the reservoirs are still widely utilized by contemporary water resources system designers. The application of the new scientific methods is particularly justified when it provides answers to questions that cannot be resolved by traditional methods – for example, in relation to the optimal sizing of different systems of reservoirs; also, the reservoir systems cannot be effectually designed without taking the risks of operation (that is based on probabilistic considerations) into account. Some examples of both deterministic and stochastic procedures employed in reservoir capacity-yield estimations are discussed in Chapter 7.

Presently there is a general tendency for the application of stochastic, analytical, and computational methods – although it must be acknowledged that some of the theoretical-based methods sometimes cannot be used by themselves in practice, especially in the case of limited and/or 'unreliable' data series, under which circumstances designers may choose a deterministic method of approach.

Notwithstanding any shortcomings, it is an undisputable fact that the stochastic methods of approach tend to offer a more complete use of the information content of the usually limited data series. The result is the increase in the variations and spectrum of the possible solutions and methods for the design and operation of complex water resources management systems.

3.5.1. DETERMINISTIC METHODS AND MODELS IN RESERVOIR DESIGN

Deterministic models are used in hydrologic design primarily as a tool in quantitatively assessing the hydrologic variables that affect the proposed design. Deterministic planning models assume that the unregulated streamflow at any site in the river basin in any time period equals the historical average, or some critical value for that location and period; this assumption obviously ignores the natural variability of such flows. Consequently, such models that are based on average or mean values of inputs are usually too optimistic. System benefits are overestimated, and costs and losses are underestimated – if they are based only on the expected values of each input variable, instead of being based on the probability distributions describing those variables. Thus, even for the preliminary identification of efficient project designs and operating policies prior to a detailed simulation study, deterministic models tend to be of limited value. It is in the light of these limitations that many stochastic planning and design models (see Chapters 5 through 8) have been developed to account for hydrologic uncertainty.

In fact, the deterministic approach to river basin planning can be said to be the classical or traditional tool preceding the development of stochastic methodologies. However, because deterministic models for river basin studies do not explicitly consider uncertainty in hydrologic variables and/or model parameters, such models provide a rather limited picture of water resources planning management problems. Nevertheless, deterministic models can be relied upon for certain preliminary studies – i.e., usually prior to a more detailed stochastic optimization or simulation study.

3.5.2. STOCHASTIC PROBLEMS IN THE DESIGN OF RESERVOIRS

The design of reservoirs must ideally take into account the stochastic aspect of the inflows and, in some respect, of the outflows as well. Stochastic methods typically come into use when the deterministic approach is found to be deficient. Even so, stochastic processes are often analyzed in such a manner, and conclusions associated with them drawn and treated as if the process was deterministic. In any case, stochastic data do not improve poor records, but merely improve the quality of designs made with whatever records are available (Fiering and Jackson, 1971) – and the processes involved provide an idea of the confidence that can be placed on the adopted design value (McMahon and Mein, 1986).

Indeed, stochastic methods are by no means a panacea for water resources planning or design. Many shortcomings, such as operational bias and lack of knowledge concerning the underlying stochastic processes, might tend to cause decisions to be less optimal than had the phenomena been treated as deterministic. Each decision that has a stochastic input, however, must be realized as such and the proper methodology employed. The use of stochastic methods in hydrology is, after all, an attempt to widen and extend our knowledge on hydrologic events and improve our decision-making ability. This is accomplished by generating longer hypothetical sequences of events based on the statistical and probability characteristics of the past records. The generated sequences of data are then used to identify the components that contribute to error and uncertainty in a proposed design – such as is the case for reservoir storage requirements.

It is noteworthy that, in the application of synthetic hydrology to reservoir systems design, the modeling of persistence merits careful consideration since this characteristic directly determines reservoir storage sizes. In general, the degree of difficulty attached to a particular stochastic method depends primarily on the complexity of the phenomena being analyzed and on the required level of accuracy of the results.

3.6. Guidelines for the Hydrological Dimensioning of Reservoirs

There are a variety of procedures for estimating the storage necessary to meet given water demands from some supply source(s); the relative merits of the methods of choice – deterministic and/or stochastic – are measured by the errors introduced into the solution of this reservoir design problem. In multi-reservoir systems analysis, the extent of the reservoir interconnection either among the storages themselves or at the point of demand almost invariably determines whether a single or a multi-reservoir analysis procedure is most appropriate. In fact, the single reservoir technique is only a very approximate guide to the capacity-yield-probability relationship of multiple reservoir systems. In general, the investigation of storage problems may be based on: empirical approaches (e.g., Mass Curve analyses and variations thereof); data generation procedures (involving simulation studies); and/or analytical/optimization techniques.

The mass-balance procedures (also called Rippl diagram) that are based on the historical hydrographic record and the expected variation of demand in time, is apparently the oldest and most commonly used technique. This is a deterministic approach that makes the *a priori* assumption that the storage that would have been adequate in the historical past will equally be adequate to meet the future needs; this is rarely true – albeit falls within acceptable error limits.

If the input is stochastic and the output either deterministic or independently stochastic, a Monte Carlo procedure can be used to estimate storages at given levels of reliabilities. The limitation of this approach hinges on the inability to generate hydrographs that are both equally likely, and also accurately reflect the extreme events to be anticipated. Where no such extreme occurrences can be deciphered from the record, and where other streams under similar climatic and geographic conditions do not display these effects, the procedure is probably more valid than the use of the deterministic analysis of the historical record only.

When the indicated inflow can adequately be represented by its frequency distribution function, the required storage can be estimated via queuing theory (Langbein, 1958 and 1959). However, for complex systems involving both serial and cross-correlation between system components, the computational burden can be rather heavy. Consequently, a Monte-Carlo procedure using the same statistic but which preserves serial- and cross-correlations, as well as correct for evaporation losses, is probably a more satisfactory approach to utilize.

Ultimately, the storage-space allocation in the hydrologic design of multipurpose reservoirs may follow a general-purpose logic diagram similar in structure to the flowchart shown in Figure 3.10.

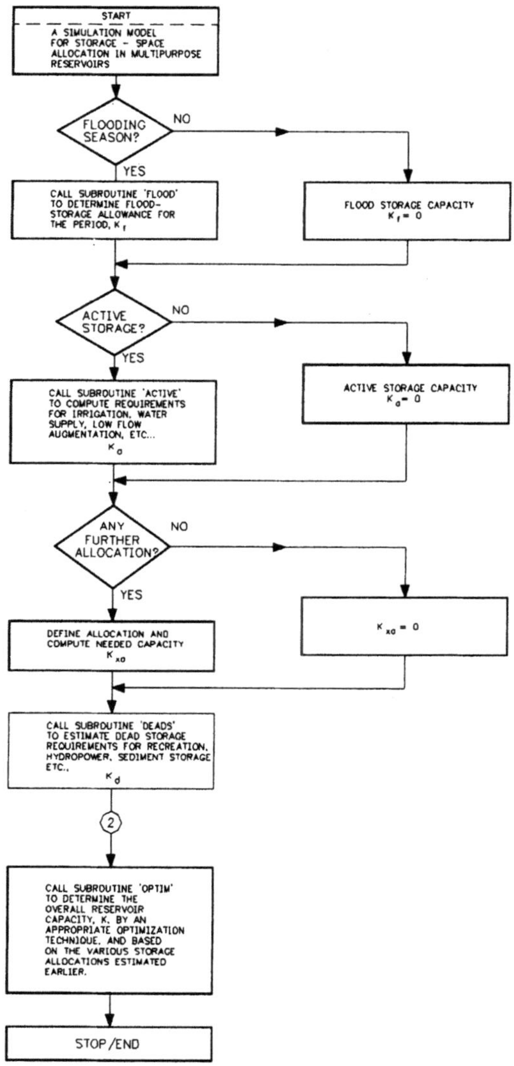

Figure 3.10. Logic diagram for the hydrologic design of multipurpose reservoirs

Chapter 4

PRINCIPLES AND CONCEPTS IN THE HYDROLOGIC DESIGN AND OPERATION OF STORAGE RESERVOIRS

This chapter discusses key fundamental principles and concepts that will help in the efficient design of water control systems – and also make it possible to develop and implement optimal reservoir design and operation management decisions. In fact, of significant, even fundamental importance in the hydrologic design and operation of water management systems are the concepts of risk and reliability. In general, regardless of the type of reservoir storage provided or its mode of operation, an element of hydrologic risk exists in all such systems due to the inherent possibility for system failures to occur. Thus, complete disregard of failure probability in the systems analysis may introduce negative elements similar to the statistical problem of the so-called 'ruin of the gambler' into the project. Indeed, it is important to include risk elements into the hydrologic design of reservoirs in order to assure the birth of a reliable system.

4.1. Utilization of Reliability-Based Techniques in the Hydrologic Design Process

The objective of storage reservoirs is the regulation and transformation of the stochastic processes of natural flows, with due consideration given to the prevailing or anticipated water resources management needs. In this effort, an important primary task involves the determination of the so-called 'efficiency function' of the reservoir; the *reservoir efficiency function* can be characterized by using:

- the maximum discharge carried to the downstream section of the river, Q_d^{max} – in the case of flood protection reservoirs; and
- the flow volume that satisfies water supply requirements, M, over time.

The efficiency functions associated with these alternative reservoir storages are elaborated below. In any case, it should be noted here that, further to the above-indicated storage requirements, the relevant parameters associated with energy production, fishery, wildlife protection and recreation, among others, should also be given due consideration in the reservoir design process.

4.1.1. THE CONCEPT OF RESERVOIR EFFICIENCY FUNCTIONS

The efficiency function of a flood protection reservoir can be characterized by the following relationship:

$$K = f(Q_d^{\max}) \tag{4.1}$$

where K represents the storage capacity that ensures that the outflow, Q_d^{max}, from the reservoir will be lower then the maximum upstream flood flow, Q_u^{max}, i.e.,

$$Q_d^{\max} < Q_u^{\max} \tag{4.2}$$

The general nature of this efficiency function is illustrated in Figure 4.1 – using data from the Hasznos reservoir in Hungary as an example.

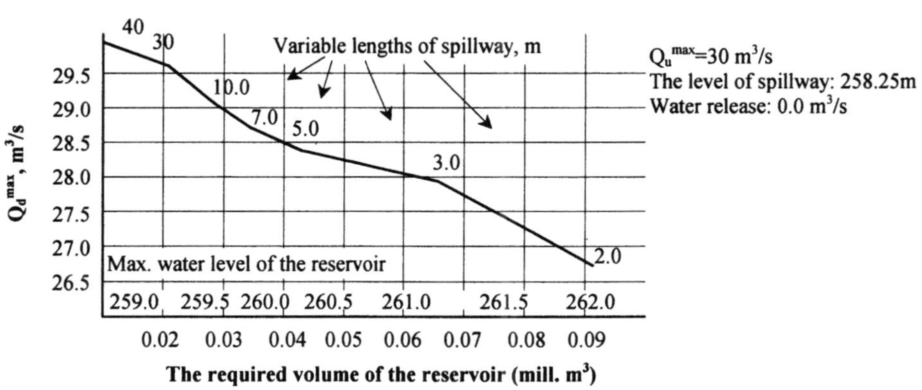

Figure 4.1. The efficiency function for a flood retention reservoir

It is obvious that a (imaginary) 'reservoir' having a volume of K = 0 will not modify the Q_u^{max} value – so the starting point of the efficiency function curve is:

$$K = 0, \quad \text{and} \quad Q_d^{max} = Q_u^{max} \tag{4.3}$$

If the volume of the reservoir is set to K = V, where V is the whole volume of the flood flow, then the outflow becomes:

$$K = V, \quad \text{and} \quad Q_d^{max} = 0 \tag{4.4}$$

which will determine the last point on the efficiency function curve.

In the case of storage reservoirs used for water supply, the efficiency function is represented by the following relationship:

$$K = f(M) \tag{4.5}$$

where M is the water demand that can be satisfied by the reservoir. In this case, it is apparent that:

$$\lim_{K \to \infty} M_{max} \to \overline{V} \tag{4.6}$$

where $\overline{V}$ is the multi-annual average volume of the flow. A continuous water demand corresponds to the lowest discharge (LKQ) of the river that can be satisfied without storage – in which case the efficiency function is a monotonic function that starts at the following point:

$$K = 0, M = LKQ \tag{4.7a}$$

and approaches the value of:

$$K \to \infty, M \to \overline{V} \tag{4.7b}$$

as illustrated in the Figure 4.2.

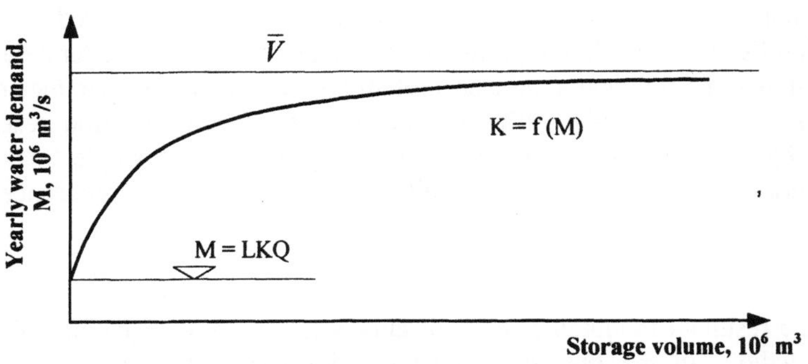

Figure 4.2. The efficiency function of the reservoirs for water supply

The efficiency functions presented above should generally be determined on the basis of the water balance equation, viz.:

$$\sum_t Q_u - \sum_t Q_d = \underset{t}{D}(S) \tag{4.8}$$

where:

$\sum_t Q_u$ = the inflow rate to the reservoir [m^3];

$\sum_t Q_d$ = the outflow rate from the reservoir [m^3]; and

$\underset{t}{D}(S)$ = the rate of change of storage (S) within the reservoir in a given time period, t.

The objective of the hydrological dimensioning is the determination of the volume (K) of the reservoir as a function of Q_d, which is the maximum of the $\underset{t}{D}(S)$ value. Applying equation (4.8) in the context of flood retention reservoirs, the extreme value of the function is given by:

$$K = \max \underset{t}{D}(S) \tag{4.9}$$

under the following condition:

$$Q_d = Q_d^{max} \tag{4.10}$$

Applying equation (4.8) in the context water supply reservoirs, the volume of the reservoir satisfying the water demand requirements is given as follows:

$$K = \max \underset{t}{D}(S) \tag{4.11}$$

under the following condition:

$$Q_d = M \tag{4.12}$$

The applied time unit should be 'seconds' in the case of flood retention, and 'month' or 'year' when the reservoir is used for conservation storage.

Recognizing that streamflows are better described by a structured stochastic process, probabilistic factors should preferably be incorporated in the determination of the efficiency function – to yield a more accurate representation of this parameter. This means that, in the case of the efficiency function for flood retention reservoirs, characterized by equation (4.1), we have to calculate not only a single 'maximum' flood, but a series of flood flows that have different probabilities of exceedance, P, as follows:

$$K = F(Q_d^{max}, P) \tag{4.13}$$

For each of the resultant monotonic series of curves (Figure 4.3), equations (4.3) and (4.4) remain valid. In the case of storage reservoirs used for water supply, the efficiency function is represented by:

$$K = f(M, P) \tag{4.14}$$

where P is the reliability of supply by the storage reservoir of volume K that meets the water demand.

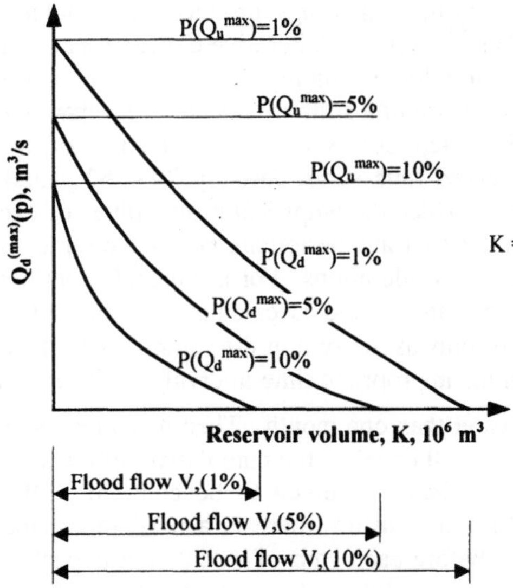

Figure 4.3. Efficiency function of a flood retention reservoir (with stochastic inputs)

4.1.2. THE EFFICIENCY FUNCTION AS A BASIS FOR STORAGE DETERMINATION

One possible approach to the dimensioning of storage reservoirs involves the determination of the efficiency function, defined by the following general relationship:

$$K = f(M, P) \tag{4.15}$$

where K is the volume of the reservoir satisfying the needs, M, with given reliability, P. For the solution of this problem in the case of a single reservoir, one can use the continuity equation, viz.:

$$Q_u(t)\Delta t - Q_d(t)\Delta t = \Delta S(t) \tag{4.16}$$

where $Q_u(t)$ is the volume of water entering the reservoir in time period *t*; $Q_d(t) = M(t)$ is the outflow (water supply); and $\Delta S(t)$ is the change in water volume in the reservoir during time interval Δt.

The task of the dimensioning of the reservoir consists of the determination of the necessary volume change, $\Delta S(t)$, that satisfies the water supply needs, M(t), during the expected life time of the reservoir, ($T = t_x - t_o$). Consequently the time interval, Δt, should be determined as a function of the inflow and the design needs. If Δh_i is the time period of water shortage, then the task becomes one of determining the following value:

$$\max\left\{ \int_{\Delta h_i} Q_u(t)dt - \int_{\Delta h_i} M(t) \right\} = \Delta S(t) \tag{4.17}$$

Equation (4.17) is more of a theoretical representation of the problem; any practical application depends on the exactness and reliability of the forecast of the inflows and water demands. In a typical engineering design practice, the problem usually can be approximately solved on the basis of repeated, iterative use of observed data.

Overall, the simplest method for estimating the volume of a reservoir that can operate satisfactorily involves the use of a suitably selected year from the past period of observations. Thus, the first step consists of the selection of a 'characteristic' hydrologic year from the observed time series (usually 20 – 50 years) of the inflow. The characteristic year is that in which the empirical probabilities of exceedance of the sum of annual flow and the water shortage period are nearly the same – and this can be selected on the basis of physical considerations. For instance, in continental conditions, the sum of the flow of the shortage period is calculated as the sum of flow of the summer and autumn periods, or only as the sum of the summer period. The next step of the calculation is to decide on the appropriate time interval, Δt; for example, in the case of $K \geq 0.3\,\overline{V}$, this is set at greater than one month. Then it can be assumed that, within the Δt time period, the reservoir will equalize the cumulative inflow and the withdrawal – with the values of Q_u and M being assumed to be constant. Of course, for the determination of the $K = f(M)$ relationship, there is a need to calculate different M values. This also means that, before each step of calculation is performed, there is a need to fix the $M = Q \cdot D_t$ water withdrawal (where Q is the discharge for water demand). The value of Q is usually constant in the case of D_t = one year, but it can be different when D_t = one month, and/or if the reservoir is used for irrigation purposes. The calculation should be started with the values of $Q_u(t)$ observed in the first Δt_1 time period of shortage, and then eventually arriving at the following relationship:

$$Q_u(t)\Delta t_1 - Q_d(t)\Delta t_1 = \Delta S_1 \tag{4.18}$$

Next, we continue the calculation and summation of the ΔS differences (starting from the first negative values of ΔS) until such point in time when the sum reaches a positive value. The subsequent step is the summation (starting from the next negative difference) of the values, until a positive value is reached once again. The minimum value of the negative sums equals the volume necessary for the study reservoir.

For the above-indicated procedure, it is assumed that the reservoir is full before the operation, and that the draw-off is smaller than the multi-annual average discharge. To determine the efficiency function, $K = f(M)$, it becomes necessary to carry out the calculations for different $M = Q \cdot D_t$ values – usually achievable by using an appropriate computer algorithm. The results obtained from an application of this procedure to the Borjad reservoir in Hungary is shown in Table 4.1; in this case, the value of the storage volume is determined to be $\Delta S_{min} = -15.9 \cdot 10^6$ m^3 for a characteristic year.

It is noteworthy that, the above-indicated behavior analysis for estimating storage size has a number of limitations. Indeed, the assumption that the reservoir is initially full can affect the probability of failure if low flows occur at the beginning of the historical record because the sequencing of inflows may not be representative of the population of flows. An alternative approach proposed by Klemes (1993) is to use the final reservoir level as the starting level for a second sequence where the observed data are used in the same sequence twice, end-to-end. This ensures a water balance over the period of record (Maidment, 1993).

Table 4.1. The Basic Task of Hydrologic Design

		Characteristic hydrologic year											
Month		XI	XII	I	II	III	IV	V	VI	VII	VIII	IX	X
Water demand	Q, m^3/s	6.4	6.4	6.4	6.4	6.4	6.4	6.4	6.4	6.4	6.4	6.4	6.4
	V=M,10^6 m^3	16.6	17.1	17.1	15.2	17.1	16.6	17.1	16.6	17.1	17.1	16.6	17.1
Inflow	Q, m^3/s	4.1	4.6	4.9	11.1	5.4	4.8	4.2	13.9	14.0	3.2	3.6	3.5
	V, 10^6 m^3	10.6	12.3	13.1	26.9	14.4	12.4	11.2	36.1	37.4	8.6	9.3	9.2
Change of the water volume	V, 10^6 m^3	-6.0	-4.8	-4.0	11.7	-2.7	-4.2	-5.9	19.5	20.2	-8.5	-7.3	-7.9
Water balances	Water demand, M	-6.0	-4.8	-4.0	11.7	-2.7	-4.2	-5.9	19.5	20.2	-8.5	-7.3	-7.9
	Inflow, V	16.6	33.7	50.8	66.0	83.1	99.7	116.8	133.4	150.5	167.6	184.2	201.3
	Differences, S	10.6	22.9	36.0	62.9	77.3	89.7	100.9	137.0	174.4	183.0	192.3	201.5

4.1.3. GENERAL TYPES OF RELIABILITY PARAMETERS VS. RESERVOIR EFFICIENCY FUNCTIONS

Reliability is a measure of the level of dependability at which project needs and requirements can be met; it is the complement of the risk of failure, or the probability that the pre-specified demand/condition will not be met. Reservoir reliability is an expression of the probability that a specified demand or condition will be met in a given future time period. Indeed, various definitions of reliability can be formulated to serve the purposes of the particular study. In particular, two parameter types may be used to characterize reliability – namely, *temporal reliability* and *supply reliability*; these are elaborated below.

- *Temporal reliability.* The temporal reliability of a storage reservoir is defined as the proportion of time that the reservoir system is able to meet demand, represented by the following general relationship:

$$P_t = \lim_{n \to \infty} \frac{m}{n} \tag{4.19}$$

where m is the number of the years when the water supply requirement, M, is satisfied and n is the number of years in the investigated time period. The complementary expression of the temporal reliability or safety is the risk of failure (to satisfy the water demand, M), represented as follows:

$$P_0 = 1 - P_t \tag{4.20}$$

which is the probability of emptying of the reservoir during the operation.

The reservoir efficiency function provided in equation (4.15) – containing three variables – corresponds to a series of conditional probability distribution functions, viz.:

$$F(x|y) = P(M \leq x|K = y) \tag{4.21}$$

This represents the probability of the situation when the water use is less than or equal to an optional value of x, given the condition that the volume (or capacity) of the reservoir, K, equals y; the domain of values of the function, F(x|y), is as follows:

$$0 \leq F(x|y) \leq 1 \tag{4.22}$$

It is noteworthy that equation (4.15) is the extension of equation (4.5). It is apparent that the values of temporal safety, P_t, will increase continuously in the case of identical demands and supply (i.e., for $M = x$ and $K = y$). Indeed, the larger the volume of the reservoir, the greater is the probability of satisfying the water supply needs, M. Thus,

$$P_t(M \leq x_i|K = K_1) < P_t(M \leq x|K = K_2) \quad \text{if } K_1 < K_2 \tag{4.23}$$

If there is a fixed value of K = *y*, the reliability, P_t, will decrease in the case of an increasing M value. This means that, for a reservoir of a given capacity, it is possible to satisfy an increased water demand only with lowered reliability (illustrated by an example in Figure 4.4). Consequently,

$$P_t(M \leq M_1|K = K_i) < P_t(M \leq M_2|K = K_i) \qquad \text{if } M_1 < M_2 \qquad (4.24)$$

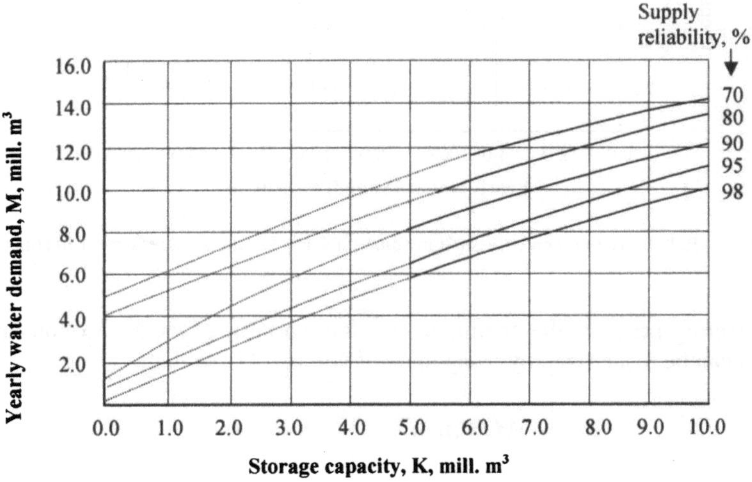

Figure 4.4. The efficiency function, $K=f(M,P_t)$, for the Borjad reservoir in Hungary

The modified equation (4.15) can indeed be presented by a series of conditional probability distribution functions, identified by the K = *y* parameter. The K = *y* parameters of the continuously increasing distribution curves, representing the risk or probability of exceedance of water supply of $M \leq x$, will have a decreasing character. It is obvious that, in the case of an identical M = *x* value, by increasing the reservoir capacity, K = *y*, the risk of failure or probability of break down of the water supply, P_0, will decrease – and correspondingly, there is an increase of the probability, $P_t = (1 - P_0)$, that represents the temporal reliability of the water supply (illustrated by an example in Figure 4.5).

- *Supply reliability*. In the demand-supply world, it is possible to have a situation where there is depletion of the reservoir, or where there are intermittent interruptions of the water supply in a given year – but this does not mean that there is no water supply at all. In the affected years, there is a possibility to ensure that a certain portion, $M_r < M$, of the total water demand, M, will be available for the water supply. Under such circumstances, the quantitative reservoir safety or *supply reliability*, P_v, of water supply from a reservoir of capacity K, can be defined as the proportion of expected values $\overline{M_r}$ to M, viz.:

$$P_v = \frac{\overline{M_r}}{M} \qquad (4.25)$$

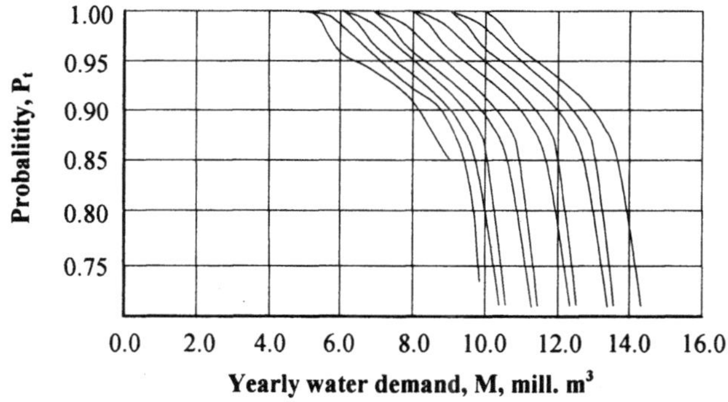

Figure 4.5. Reliability function curves (from equation 4.15) for the Borjad reservoir in Hungary

In engineering practice, the continuous random variable of the M = x consumption scenario can be discretized by using units of volume, D_v, so that:

$$M = m \cdot D_v \qquad m = 0, 1, 2, \ldots \tag{4.26}$$

Consequently, the expected value of the water volume actually delivered is given by:

$$\overline{M}_r = \sum_{i=0}^{m} i \cdot D_v \cdot P_i \tag{4.27}$$

The probability of being able to meet the water demand requirements of $M_r = M = m \cdot D_v$ is equivalent to the temporal reliability, P_t. If we are able to calculate the P_i belonging to each of the values of m_r, i.e.,

$$m_r = i \cdot D_v \qquad i = 1, 2, \ldots (m-1) \tag{4.28}$$

then the modified equation (4.27) can be represented by the following relationship:

$$\overline{M}_r = \sum_{i=0}^{m-1} i \cdot D_v \cdot P_i + m \cdot D_v \cdot P_t \tag{4.29}$$

It is apparent that the quantitative supply reliability, P_v, is not a probabilistic value – and can be defined as follows:

$$P_v = \frac{\overline{M}_r}{M} = \frac{\sum_{i=0}^{m-1} i \cdot D_v \cdot P_i + m \cdot D_v \cdot P_t}{m \cdot D_v} = \frac{\sum_{i=0}^{m-1} i \cdot D_v \cdot P_i}{m \cdot D_v} + P_t \tag{4.30}$$

Considering the fact that the *i* values, the D_v unit volumes, and the P_i probabilities are nonnegative values, it becomes evident that the supply reliability cannot be lower than the temporal reliability, i.e.,

$$0 \leq P_t \leq P_v \leq 1 \tag{4.31}$$

It should be mentioned again that the supply reliability, P_v, is not a probabilistic value – because it is simply a proportion defined by equation (4.25). The corresponding series of the reservoir efficiency function curves with given P_v values can be presented in a system of coordinates, where the M-values are on the vertical scale and the reservoir volumes, $K = k \cdot D_v$ are fitted on the horizontal axis – as illustrated by Figure 4.6.

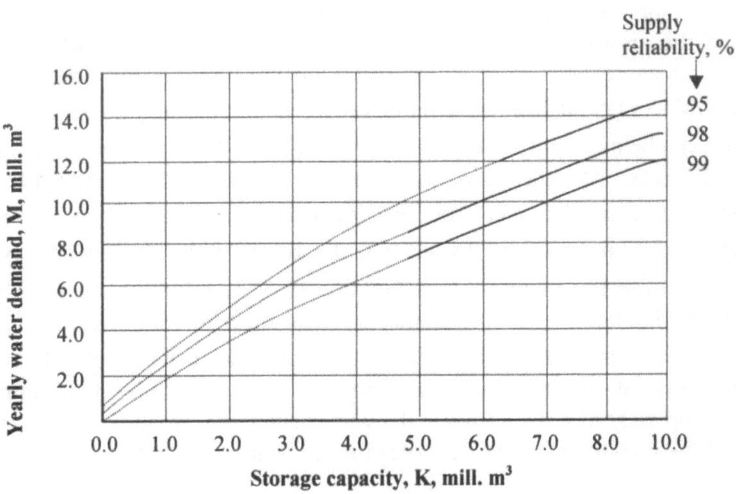

Figure 4.6. Efficiency function, [K=f (M,P_v)], characterized by the supply reliability, P_v for the Borjad reservoir in Hungary

Reliability estimates are typically developed from the results of a reservoir system simulation.

4.1.4. STORAGE ALLOCATION IN MULTIPURPOSE RESERVOIRS

Different reliabilities may be required when releasing water from a reservoir to meet demands associated various uses; for example, water for the cooling of thermal power plants must be delivered with greater reliability than water for irrigation of meadows or pastures, or for fish pond, or for recreation. Thus, for the sizing of multipurpose reservoirs, a method of analysis is presented here, in which the total water demand, *M*, is separated into several components that have different corresponding reliabilities, *p*. Indeed, different benefits and reliabilities required for various purposes can be balanced, such that the release, *M*, serving as the basis of reservoir sizing is divided up into consumptions of different reliability requirements – as follows:

$$M = M(1) \cdot p(1) + M(2) \cdot p(2) + \ldots + M(n) \cdot p(n) \qquad (4.32)$$

where p(1) > p(2) > ... > p(n-1) > p(n), with p(1) for the water demand M(1) tolerating the lowest risk (where P = 1-p(1)) – associated with the high reliability of supplying cooling water for a thermal power plant, and p(n) the lowest reliability at which demand can still be met for such purposes as water for irrigation of meadows and pastures. For example, in the case of the thermal power plant, p(1) may be set at ≅ 98%, and for the meadows and pastures, p(n) ≅ 70%. On such basis, probability distribution functions of these restricted releases and downstream flood discharges that correspond to various water allocation schedules can be developed – as illustrated by Figure 4.7.

Ultimately, the comprehensive water allocation schedule can be derived by successive approximations in a manner shown in Figure 4.8; the graphs A-C show the steps of the approximation process. For example, graph (B) shows the probability distribution curve associated to the restricted consumption of (industrial) firm water demand of 15·DV. It follows from this representation that an industrial water demand of 8·DV can be met with 95% reliability, 3·DV with 80%, and the remaining 4·DV with 50% reliability. In the subsequent second approximation step, the estimation is repeated for release to meet the firm demand of 8·DV, and to follow the rate of agricultural consumption of 7·DV. On the basis of the second approximation, it is possible to raise the industrial consumption of 95% reliability to 10·DV – albeit the third approximation suggests that this increase may be exaggerated. The final result from the fourth approximation is 9·DV, of which 3·DV can be met with 80% reliability. Consequently, along the downstream stream section, the hazard of summer floods is nearly completely eliminated while the winter snowmelt flood peaks are reduced to one fourth of the incoming flood discharges for the continental conditions.

4.2. Topographical Characteristic of the Reservoir

As part of the reservoir design exercise, the designer is planning not only the storage volume of the reservoir, but also the dam structure that creates the reservoir. Thus, it is simply not enough to have adequately determined the volume of the reservoir; it is equally important to adequately determine several other important features – including the water level corresponding to the design volume of the reservoir, and also the height of the dam.

In the process involved in the hydrologic dimensioning efforts, the designated volumes of the reservoir, $K = k \cdot D_v$ should be combined with the depth of water, H – such that the relationship K = *f*(H) can be presented in a (K,H) system of coordinates (Figure 4.9). Using the topographic curve of the reservoir, the characteristic curve, A = *f*(H) can also be constructed – where A is the area of the water surface of the reservoir associated with depth H of the cross-section of the dam. The following integral curve of the A = *f*(H) function represents the 'volume curve' for the reservoir (Figure 4.9):

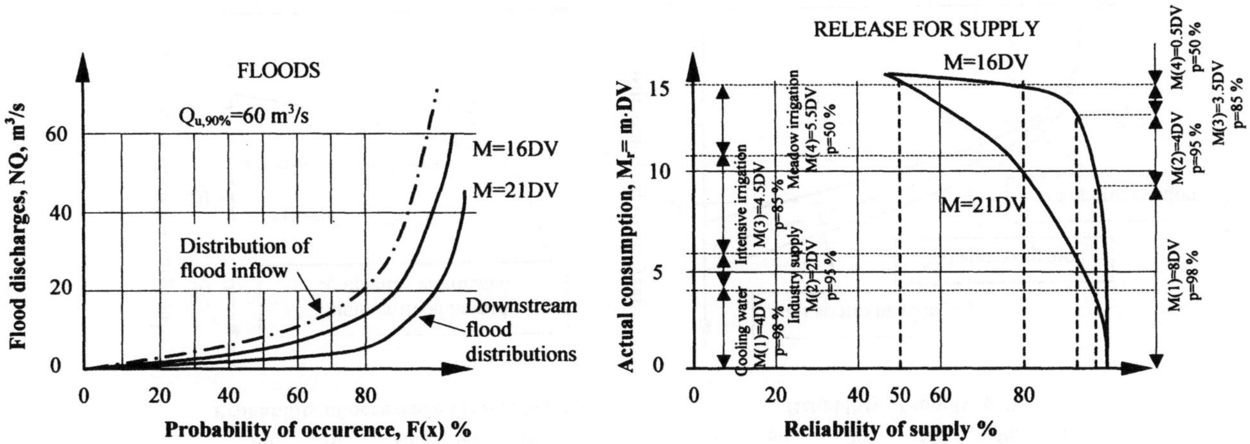

Figure 4.7. Water allocation schedule for the Gaja Creek at Fehérvárcsurgó, Hungary, with K = 16DV, and DV = 10^6 m^3 and the case of demand M = 21DV and M = 16DV

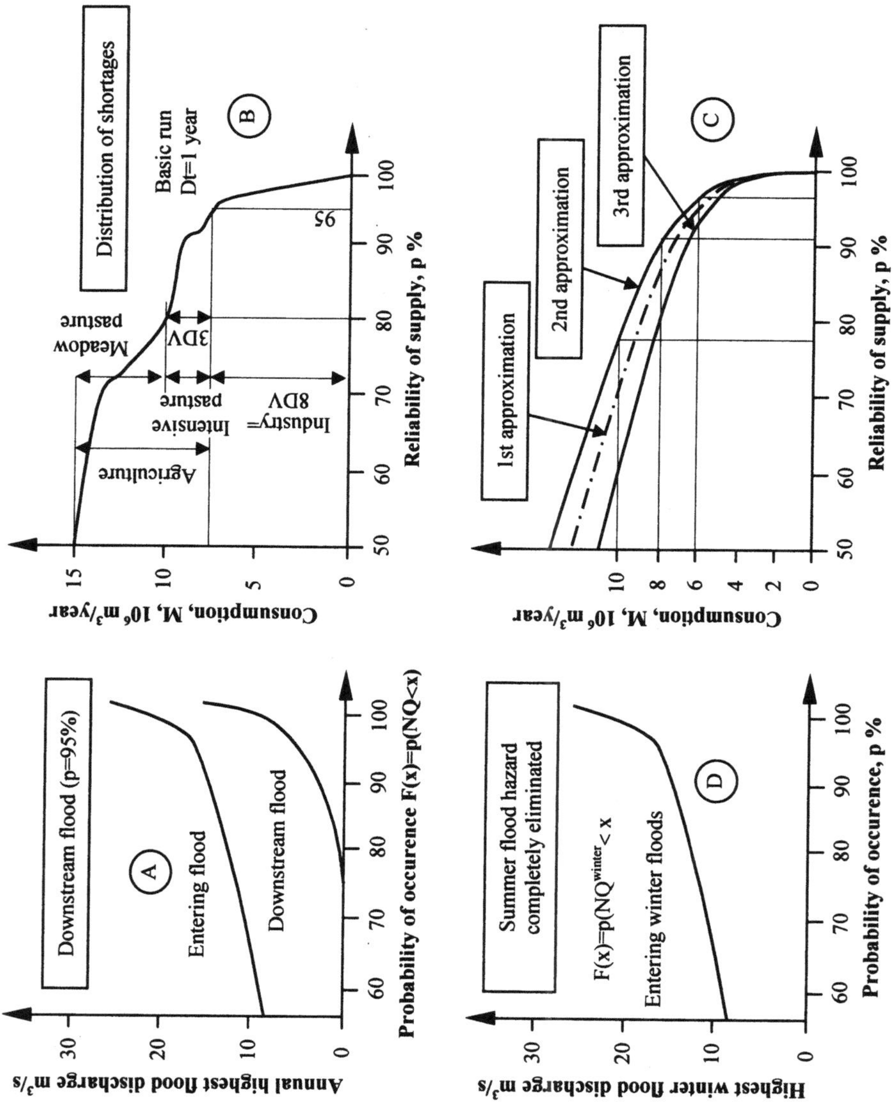

Figure 4.8. A complex study of the Borjad reservoir on the Karasica river, Hungary, with $K=20.4 \cdot 10^6$ m^3

$$K(H) = f(H) = \int_{H=0}^{H\max} [A = f(H)]dH \tag{4.33}$$

These topographical relationships can also be used in the evaluation and description of secondary processes such as seepage, evaporation, sedimentation – which are all important considerations in the reservoir design process.

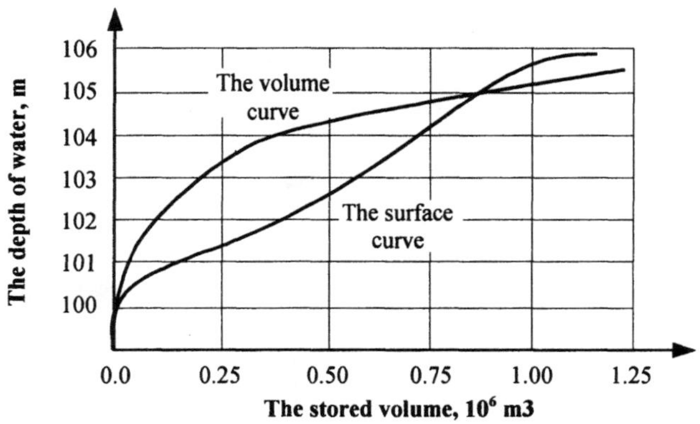

Figure 4.9. Reservoir topographic curves

4.3. Modeling Methodology for the Systems Simulation of Reservoir Design Problems

In river basin planning, the number of factors and questions to be addressed and the uncertainties of future streamflow preclude the use of a single mathematical model in the analysis of such systems. Thus, a typical methodology for such types of projects may include a screening submodel, a simulation submodel, and a sequencing submodel, as appropriate. Although such methodologically compartmentalized design cannot be expected to be as good as one designed by a single model that could address all the relevant issues, this type of sub-optimization is an almost inevitable concomitant when dealing with such complex problems as are often encountered in multireservoir systems design. Indeed, by using such submodels together in a sequential manner, reliable answers can be found to the pertinent design questions. For reservoir capacity decisions, mass balance constraints – which trace the flow of water through the river system (comprising of reservoirs for irrigation, hydropower production, diversions, etc.) – form the basis for model formulation and solution, as discussed below and further elaborated in greater detail by Major and Lenton (1979).

- *Continuity Constraints.* The continuity constraint in the model formulation ensures the conservation of mass at every point in the river at which water is stored, diverted/exported, or imported. As illustrated by Figure 4.10(a), the basic continuity equation is given by:

$$S_i(t+l) = S_i(t) + Q_i(t) + I_i(t) - E_i(t) - D_i(t) \tag{4.34}$$

where the subscripts *i* and *t* refer to the (project) site and time period, respectively; $S_i(t+l)$, $S_i(t)$ represent reservoir storage at the beginning of future/next season, *t+1*, and of the present season, respectively; $Q_i(t)$, $I_i(t)$ are inflows and imports, respectively; and $D_i(t)$, $E_i(t)$ represent the reservoir release and diversion/exports, respectively.

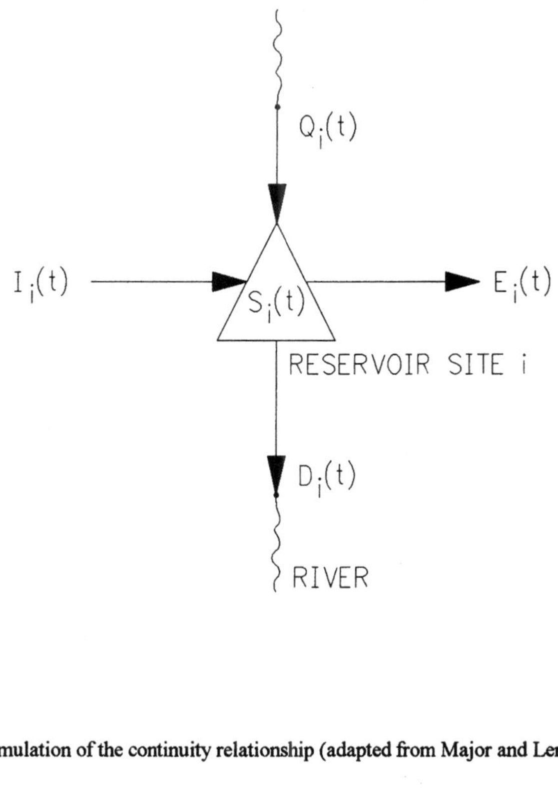

Figure 4.10(a). Formulation of the continuity relationship (adapted from Major and Lenton, 1979)

The upstream flow $Q_i(t)$ depends on natural streamflow, and on the decision made immediately upstream at site *i-1*; Figure 4.10(b) illustrates a process formulation used to express $Q_i(t)$ as a function of these two effects. Thence, using the principle of conservation of mass yields the following:

$$Q_i(t) = D_{i-1}(t) + \Delta F \tag{4.35}$$

where ΔF represents the increment to natural streamflow between sites *i-1* and *i*. Consequently,

$$S_i(t-l) - S_i(t) + D_i(t) + E_i(t) - I_i(t) - D_{i-1}(t) = \Delta F \tag{4.36}$$

For a deterministic model, ΔF is based on mean seasonal streamflows that are taken to occur with certainty; the value of ΔF is taken as the difference between the mean seasonal streamflows at sites *i-1* and *i*, viz.:

$$\Delta F = F_i(t) - F_{i-1}(t) \tag{4.37}$$

and this represents the flow into the stream from the drainage area between *i-1* and *i*. $F_i(t)$ and $F_{i-1}(t)$ are obtained from streamflow records available at gaging stations on the river. ΔF is not the total streamflow at site *i* because $D_{i-1}(t)$ is the flow in the stream at *i-1* after development of the basin. Thus, ΔF represents the change (increment or decrement) to $D_{i-1}(t)$ due to natural effects (such as runoff from precipitation and/or snowmelt, tributary flow evaporation, infiltration or exfiltration, etc.). Subsequently, the general stream continuity equation/constraint is obtained for all sites *i* and seasons *t* as follows:

$$S_i(t+l) - S_i(t) + D_i(t) + E_i(t) - I_i(t) - D_{i-1}(t) = F_i(t) - F_{i-1}(t) \tag{4.38}$$

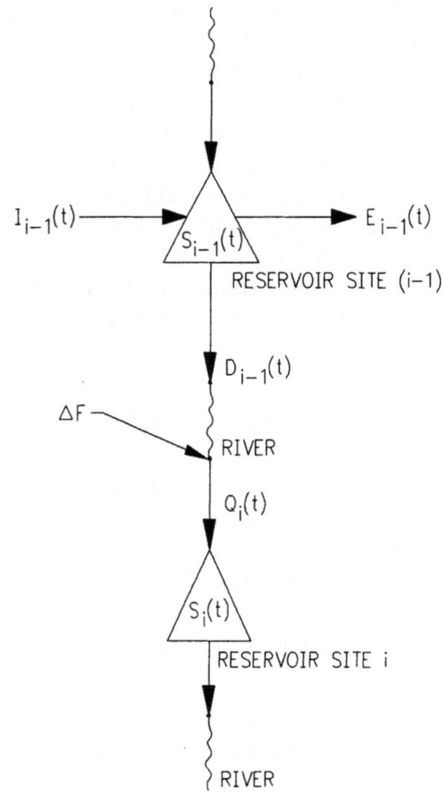

Figure 4.10(b). Further development of the continuity relationship with an upstream site (adapted from Major and Lenton, 1979)

- *Reservoir Constraints.* For any site i and during any season t, the storage in a reservoir cannot exceed the storage capacity, K_i (for site i), so that

$$S_i(t) - K_i \leq 0 \qquad \forall i,t \tag{4.39}$$

By making K_i a decision variable in the design formulation model, the optimal storage capacity of the reservoir can be determined. Furthermore, the storage-head relationship is given as follows:

$$S_i(t) - \sigma_i\{H_i(t)\} = 0 \qquad \forall i,t \tag{4.40}$$

where $\sigma_i\{H_i(t)\}$ is a function relating storage volumes to water elevations in the reservoir (i.e., height of water behind the dam). This function depends on the shape of the valley at site i.

- *Irrigation Constraints.* The irrigation process is an extremely complex one – this complexity being attributable to the great number of variables (e.g., irrigation water volume and quality; temporal distribution of irrigation water volumes; solar radiation; precipitation; crop types; various soil properties; etc.) that affect agricultural production. What is desired in modeling irrigation systems is a function relating crop yield to quantities of irrigation water supplied – i.e., the production function. Indeed, the agricultural production function has many dimensions, one for each of the variables that affect crop yield. The irrigation constraints, with respect to irrigation water requirements can be derived with reference to the typical elements shown in Figure 4.11. A constraint relating the water volume reaching the irrigation site, $IRGN_i(t)$, to the flow diverted out of the stream, $E_i(t)$, is:

$$IRGN_i(t) - [1 - \varepsilon_i(t)]E_i(t) = 0 \qquad \forall i,t \tag{4.41}$$

where $\varepsilon_i(t)$ is a coefficient that measures the water lost in transmission from the stream to the irrigation zone, and which could be assumed to return to the stream. A further constraint relates the flow diverted for irrigation, $E_i(t)$, to the flow that returns to the stream from the irrigation site, $RET_i(t)$, viz.:

$$RET_i(t) - [1 - \mu_i(t)]E_i(t) = 0 \qquad \forall i,t \tag{4.42}$$

where $\mu_i(t)$ is the total loss coefficient for irrigation; in fact, this represents a combination of the transmission losses, $\varepsilon_i(t)$, and consumptive use requirements, $\rho_i(t)$, such that:

$$\mu_i(t) = \rho_i(t)[1 - \varepsilon_i(t)] \tag{4.43}$$

when all transmission losses are assumed to return to the stream.

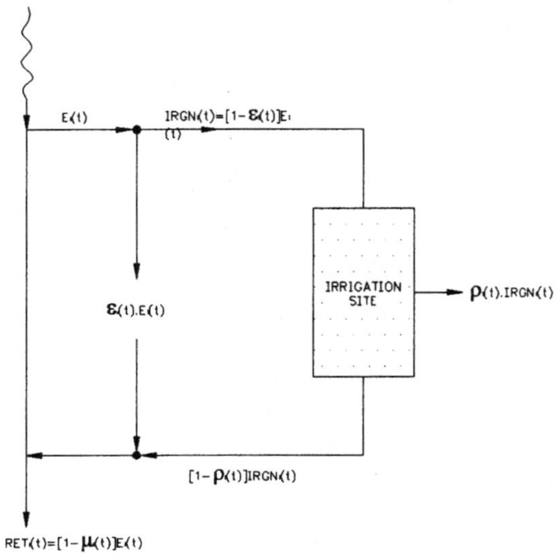

Figure 4.11. Irrigation water requirements, utilization and losses (adapted from Major and Lenton, 1979)

- *Interbasin Imports and Exports.* Interbasin transfers are modeled as diversions of water into or out of the main stream – recognizing that the periodic transfers do not exceed the capacity of the river channel. For imports,

$$I_i(t) - IMP_i(t) \leq 0 \qquad \forall i,t \tag{4.44}$$

where IMP_i is the capacity of the import canal at site *i*. For exports,

$$X_i(t) - XP_i(t) \leq 0 \qquad \forall i,t \tag{4.45}$$

where $X_i(t)$ is the average export from site *i* during period *t*, and XP_i is the capacity of the export canal at site *i*.

- *Hydropower Constraints.* There are only three primary decision variables affecting hydro-energy production – flow through the turbines of the power plant, the potential energy of flow measured by the head, and the capacity of the power plant. The production function for hydroelectric energy serves as the first constraint, viz.:

$$P_i(t) - \beta\eta D_i(t) H_i(t) \leq 0 \qquad \forall i,t \tag{4.46}$$

where $P_i(t)$ is the energy produced at site i during season t [MWh]; $H_i(t)$ is the head of water in reservoir i at beginning of period t; η is the power plant efficiency; and β is a unit conversion factor. This relation is nonlinear and non-separable because the flow through turbines, $D_i(t)$, and $H_i(t)$ – both decision variables – are multiplied together. Next, the power plant capacity represents an obvious upper bound on energy production, so that:

$$P_i(t) - V_i h(t) \leq 0 \tag{4.47}$$

where V_i is the power plant capacity in megawatts [MW]; and h(t) is number of hours in period t. This equation will be binding only if the plant produces at full capacity all the time – an unrealistic and even undesirable notion. Consequently, a load factor $\gamma_i(t)$ – defined by the ratio of the average daily production to the daily peak production – can be introduced to account for the daily variation in production. Assuming the production does not necessarily vary significantly from day to day, but rather on a seasonal level (since $P_i(t)$ is the seasonal energy production), then

$$P_i(t) - \gamma_i(t) h(t) V_i \leq 0 \quad \forall i,t \tag{4.48}$$

4.3.1. A FLOOD STORAGE SUBMODEL

The overall objective for flood protection is to select the optimal combination of reservoir capacities for a specified level of protection. The optimum solution will be the one that results in the lowest cost combination of reservoirs, viz.:

$$\min\left\{ Z = \sum_{i=1}^{n} f_i(V_i) + \sum_{i=1}^{n} g\ (W_i) \right\} \tag{4.49}$$

where $f_i(V_i)$ is the annual cost of providing flood storage, V_i, at reservoir site i $\{i = 1,2.......n\}$; and $g(W_i)$ is the annual cost of providing a spillway outlet capacity, W_i, at reservoir i. This objective is expected to be attainable, subject to a set of constraints – including storage constraints, reservoir release constraints, conservation of mass, and 'non-negativity' constraints. Each reservoir is characterized by its maximum flood storage capacity, V_i^{max}, so that, at the start of any time period, the amount $S_i(t)$ of water stored in the reservoir must not exceed the maximum available flood storage volume, viz.:

$$V_i^{\max} \geq S_i(t) \qquad \forall i,t \tag{4.50}$$

Reservoir releases are limited to the spillway discharge capacities, which in turn are related to reservoir storage volumes. In general, this is a nonlinear function, $\psi_i\{S_i(t)\}$, that may be approximated by a linear or piecewise linear function. The nonlinear form of the constraint is is given by:

$$D_i(t) \leq \psi_i\{S_i(t)\} \leq W_i \qquad \forall i,t \tag{4.51}$$

where $D_i(t)$ is the controlled outflow from reservoir *i* to the downstream reach at the start of time period *t*. The 'non-negativity' constraint requires that the reservoir storages, inflows, outflows, and river flows be nonnegative, viz.:

$$S_i(t),\ Q_i(t),\ D_i(t) \geq 0 \qquad \forall i,t \tag{4.52}$$

In a multiobjective optimization, other constraints, such as those discussed earlier on (e.g., for irrigation, hydropower, etc.) should be given due consideration in the overall model formulation.

4.4. Optimal Release Policies in the Operation of Multipurpose Reservoirs

The optimum design of any system is dependent not only on the objectives and related specifications, but also on the probable operating policy adopted by policymakers. The analysis of operating policies for optimal output of a given reservoir system can therefore serve as the logical basis for developing the optimal design. In fact, operating policy is a necessary component of every reservoir simulation model.

The main feature of a good management scheme that prevents (or at least minimizes) failures in the operation of multipurpose reservoirs is to give a whole range of possible decisions that guarantees the efficient performance of the system. This allows the systems manager to choose the release policy that best integrates and synchronizes any additional information or forecasts that might have become available, as well as to further take into consideration secondary objectives that were not envisaged and therefore not included in the initial formulation of the problem. Invariably, reservoir operation is aimed at avoiding extreme and unacceptable failures to meet the objectives when the system is under stress – reference usually being made to the 'critical periods' (namely, the driest season, the wettest season, the highest recorded flood, etc.).

Conceptually, a *minimum release policy* would guarantee the satisfaction of all demands – or at least to acceptable levels of reliability; a *maximum release policy*, on the other hand, is needed for the evaluation of flood protection capability of the reservoirs. In multipurpose reservoirs, the goal is to guarantee specified values of both objectives of satisfying demands and flood protection, concurrently; the necessary operating rules may be established by a deterministic ('MiniMax') approach, or via the more classical stochastic approach (Mohammadi and Marino, 1984; Loucks *et al.*, 1981). The time sequence of optimal releases should lead to the most effective operation of multi-reservoirs. However, the 'optimality' of decisions based on the reservoir releases is also affected to a large extent by the forecast of future inflows to the system. The set of optimal releases (or equivalently final storage volumes) in the *t*-th time period is defined by taking into account the water management situation in all future time periods which may influence the decision to be taken at the beginning of the *t*-th time period. Thus, if the optimization process is not spread over the complete simulation period, the final storage will become depleted after satisfying demands in the *t*-th period with no hedging against future requirements. The final storage from one period generally becomes the initial storage for the next period, and for steady-state conditions, the initial storage V_0 at the start of the simulation (t = 0) can be set equal to the final storage, V_T at the end of simulation period (t = T).

4.4.1. STOCHASTIC APPROACH TO ESTABLISHING AN OPTIMAL RELEASE POLICY

Often, in water resources planning, continuous stochastic processes are approximated by Markov Chains to facilitate the construction of simple stochastic optimization models. In general, the reservoir storage at any time period is constrained between a pre-set minimum, V_{min}, and maximum, V_{max}. Furthermore, it is frequently the case that the value of S(t+l) is not independent of S(t) – where S(t) and S(t+l) are the storage volumes in time periods t and $t+1$, respectively; such dependence can be modeled by a Markov Chain. This requires specification of the transition probabilities, viz.:

$$P_{ij} = \Pr\{S(t+1) = V_j / S(t) = V_i\} \tag{4.53}$$

The transition probability is the conditional probability that the next state is V_j, given that the current state is V_i ; these transition probabilities satisfy the condition that:

$$\sum_{j=1}^{n} P_{ij} = 1 \qquad \forall i \tag{4.54}$$

The transition probability characteristic of the Markovian chain can be obtained for all states; for instance, Prekopa (as cited in Zsuffa and Galai, 1987; Zsuffa, 1982) presents a series of relationships by which the transition probabilities, P_{ij}, can be found from the probabilities P_i of the inflow Q, i.e.,

$$P_i = \Pr\{\text{Inflow} = Q\} \tag{4.55}$$

The P_{ij} values are then summarized in a transition probability matrix. The computed transition probabilities can serve as the basis for defining the probability that the storage at any time stays within the constrained values of V_{min} and V_{max}, i.e.,

$$V_{min} \le S(t) \le V_{max} \tag{4.56}$$

or alternatively, used to define the failure probability of when $S(t) > V_{max}$ or $S(t) < V_{min}$. Hence, the probabilities of 'deficit' (P_d) and 'excess' (P_e) can, respectively, be defined by the following:

$$P_d = \Pr\{S(t+1)/S(t)\}\Pr\{V_{\min}/S(t+1)\} = \Pr\{V_j/V_i\}\Pr\{V_{\min}/V_j\} \tag{4.57}$$

$$P_e = \Pr\{S(t+1)/S(t)\}\Pr\{V_{\max}/S(t+1)\} = \Pr\{V_j/V_i\}\Pr\{V_{\max}/V_j\} \tag{4.58}$$

A major issue that needs to be given careful consideration here relates to the real possibilities (or otherwise) of being able to obtain a reliable forecast of the inflows. For instance, if the lead-time of the reliable forecast is only one period ahead (e.g., one month), then for each of the periods into which the year is divided (t = 1,.....T), there is:

- a vector of input data [x_t] (namely, initial storage volumes and forecast inflows for a given time interval);
- a vector of corresponding output data [Y_t] (namely, final storage volumes); and
- a vector of unknown random disturbances [z_t] (namely, reservoir inflows in the next T-t future periods).

All these fit into a common input-output relationship that is probabilistic in nature because of the random disturbances – expressed for each season by

$$[Y_t] = [x_t] + [z_t] \tag{4.59}$$

This set of equations would constitute the operation rules that secure long-term optimality of the reservoirs' operation.

4.5. The Value of Hydrologic Information in the Management of Reservoirs

Reservoir operation generally requires rules to indicate the amount of water that should be held on a month-by-month basis throughout the year. Reservoir operation rules are guides for those responsible for reservoir operation, and they apply to reservoirs being operated in a steady-state condition. In a system of reservoirs, the operating rules must also allocate purpose-specific storage spaces for each reservoir.

A good reservoir operation policy can only be based on a reliable up-to-date information source – which usually means a good data collection network and reliable hydrological forecasts. In general, the operating rule will be a function of the information currently available – albeit different information is needed for different parts of a water resource system and for different modes of system operation. In any case, reliable forecasts are necessary for the efficient operation of reservoir systems.

To obtain reliable forecasts, one must be able to successfully design an appropriate real-time forecasting model that is problem-specific. Hydrological forecasting is the prior estimate of future states of hydrological phenomena in real time. In the historical past, most hydrological forecasts were based solely on simple empirical relationships; in more recent times, however, forecasting has become much more comprehensive with networks continuously updated and automated, and with models based on more complete physical descriptions of hydrological and hydraulic processes. In fact, the power of many streamflow forecasting models lie in their ability to update their probabilistic bases continuously.

On-line hydrologic information systems for the operation of multipurpose reservoirs require very sophisticated equipment, which imply large investment, etc. The commitment of substantial funds to these systems can only be justified if there are objective reasons to believe that their high costs will be more than compensated by the increased benefits accruing from operating the system with extra information on the occurred hydrologic sequences and their short-term forecasts. Indeed, on-line hydrologic information is particularly relevant for short-term analyses that are useful to the study of rapid hydrologic phenomena like floods – and hence its value should be estimated in terms of its use for forecasting periods of high flows, as well as for the on-line control of the reservoir during such critical periods. In addition, on-line hydrologic data can also be of use in the prediction of medium- and long-term events like persistent dry conditions. Full use of short-term forecasts on upstream inflows can be done not for the operation of the reservoir during a high flow period, but to develop an adaptive control of the flood detention storage. It must be acknowledged, however, that forecasting is only a means to aid the decision-maker but not an end in itself.

Overall, optimal or near-optimal decisions can generally be arrived at by basing the systems analysis on appropriately selected information obtainable from all available data, and then basing the forecasts on these data. Indeed, due to uncertainties in the time series used for hydrological model forecasts, an exact forecast can hardly be given by even the best of algorithms. This implies that the forecast itself is nothing more than a probability distribution of the predicted variable at the specified lead-time – with the uncertainty in the forecasts generally increasing with the lead-time.

Although the design and operation of a water resource project may often be improved by the acquisition of more data, or by the judicious use of the currently available data, there are situations – as indicated by Moss *et al.* (1978) and Matalas and Jacobs (1964) – where the acquisition and the use of additional data and/or information is of minimal usefulness, or could even prove counterproductive. In measuring the worth of hydrological data, the value of any additional data can be taken as the incremental change expected in payoff or loss. In general, a well-designed process should be used for the acquisition of the requisite information, and also in making decisions in response to this information (Anderson and Burt, 1985).

Chapter 5

SYSTEMS APPROACH IN THE HYDROLOGIC DESIGN AND OPERATION OF STORAGE RESERVOIRS

Systems analysis in water resources offers a philosophical approach to studying the interdependence within the hydrologic, environmental, and economic elements under one umbrella – with an ultimate goal of improving system performance. The systems approach is generally defined as an orderly way of looking at all angles of a complex problem, with emphasis on the interrelationship rather than the analysis of separate parts of a given phenomenon (Albertson *et al.* 1971).

The systems approach of problem solution has developed from the fields of operations research and systems engineering. Operations research and systems analyses are indeed powerful tools to be used to guide the decision maker – albeit this should not be blindly relied upon, and neither should these tools be allowed to sub-plant the intuitive judgment of the decision maker; rather, they should be used in conjunction with the pertinent judgmental inputs.

Overall, the systems approach involves the application of scientific methods (e.g., mathematical models) to the solution of problems that arise from the needs and wants of a society. The main steps to be followed in the general utilization of systems procedures are elaborated more extensively elsewhere in the literature (e.g., Mays, 1996; Brotchie *et al.*, 1973). This chapter focuses on those aspects of systems analysis as applied in the modeling of hydrological processes – and that are necessary to facilitate the effectual design and operation of storage reservoirs.

5.1. Hydrologic Models in Water Resource Systems

Models of water resources systems should generally describe the essential features of such systems and the interdependencies among system components in the detail needed for the particular decisions being considered (Mays, 1996). For instance, models are used to evaluate potential future outcomes if certain actions or decisions or policies were to be implemented. Indeed, there are many kinds of water resources models that can be used to address various planning, management, and operations issues – with most of the models designed to handle or examine different levels of study details and/or different degrees of complexities and sophistication.

A hydrologic model is a simplified representation of the natural hydrologic system, and such models can be subdivided or categorized in a number of different ways (Mays, 1996; Salas *et al.*, 1980; Dooge, 1977; Clarke, 1973; Chow, 1964). In general, all models fall into one of the following three broad categories:

1. Iconic or Physical;
2. Analog; and
3. Symbolic or Mathematical.

Models may also be classified as being either *optimization* or *simulation* models; *deterministic* or *stochastic (probabilistic)* models; *static* or *dynamic* models; *investment-and-operating-policy* or strictly *operating-policy* models; *conceptual* or *empirical* models. All model types have their strengths as well as their limitations. In fact, there is no one optimal model type for all water resources management problems. Furthermore, models do not fit neatly into the deterministic, stochastic, conceptual, etc. categories – but rather each tends to contain aspects of some or all of these classes.

Systems analysts tend to exhibit a preference for utilizing mathematical models. Mathematical models – which aim at representing real physical systems in a quantitative form – have indeed become important tools in the design, synthesis/analysis, operation, and control of complex systems. In many systems studies, a number of mathematical modeling techniques may be employed to analyze different aspects or components of the same problem. The major challenge to the systems analyst is to uncover and use that particular mathematical model or set of models that best represents the system to be analyzed; Figure 5.1 provides a possible methodology that can help facilitate the process of making an objective choice amongst several mathematical models (Mays, 1996; Salas *et al.*, 1980; Dooge, 1977; Box and Jenkins, 1976). In general, a high degree of stochastic relationship between the prediction by the model and what would actually happen in the real world is required if the model is to become useful – and of course there should also be a possibility to implement appropriate modification(s) of the model as and when necessary.

In the analysis of complex water resource systems, it often is useful to formulate an overall model in terms of a hierarchy of related models. In some situations, a lower-order model will be completely embedded in or nested in a higher-order model, whereas in other instances the separate lower-order models are linked in various fashions. It is also noteworthy that, hydrological modeling has been widely associated with geographical information systems (GIS) for applications in water resources systems planning and management (e.g., Goodchild *et al.*, 1996). Indeed, GIS-based hydrological modeling may be considered a particularly important focus area for supporting reservoir design and operation studies. These applications are founded predominantly on grid cell digital evaluation models. Ultimately, the 'taste and fondness for the familiar' tends to affect the choice of hydrologic and related water resources system models. In the case of water resources systems models, for instance, the type of water use (e.g., for water supply, flood control, etc.) for which the system is being designed may form the basis of model classification and choice.

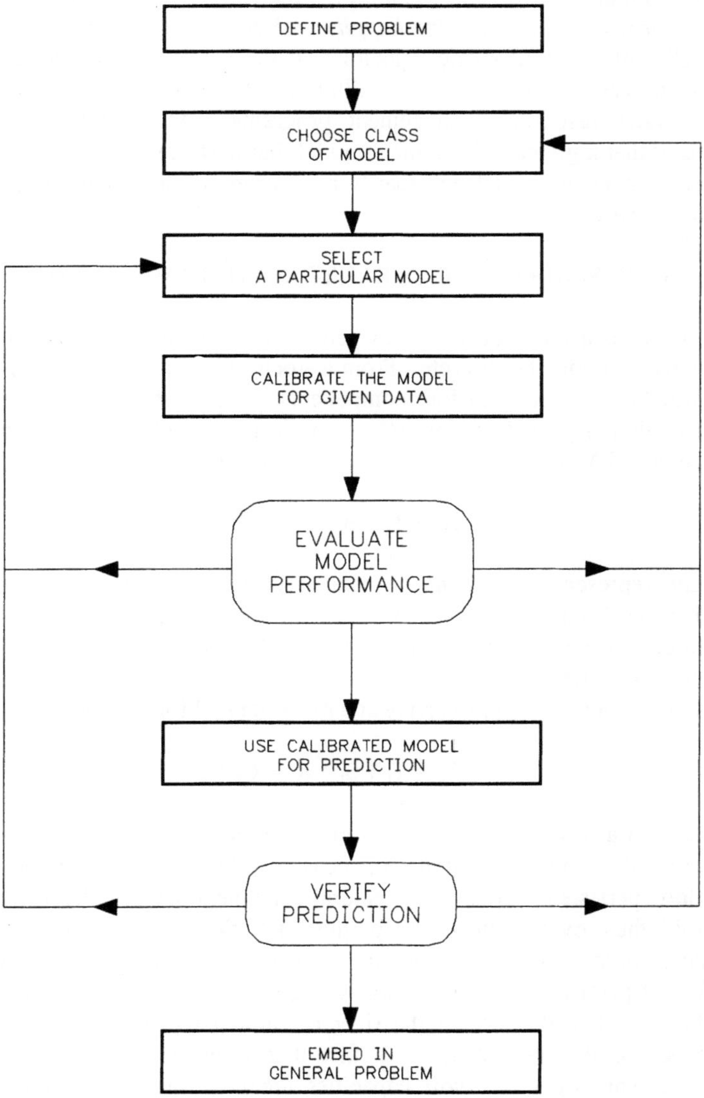

Figure 5.1. A procedure for the selection, calibration and use of a mathematical model (adapted from Dooge, 1977)

Finally, a word of caution should probably be made about mathematical models – best illustrated with the analogous view held by society about 'social' models, as expressed by Kaplan (Kaplan, 1964 – cited in Aris, 1994), that: *'Models are undeniably beautiful, and a man may justly be proud to be seen in their company. But they may have their hidden vices. The question is, after all, not only whether they are good to look at, but whether we can live happily with them.'* This view does indeed compare very well with the underlying principles in the selection and use of hydrologic and water resources system models – calling for the careful choice of such models to support water resources management programs. Most importantly, it should be recognized that a given mathematical model that performs extremely well under one set of circumstances might not necessarily be appropriate for other similar or comparable situations for a variety of reasons.

5.1.1. TIME SERIES MODELS IN HYDROLOGIC MODELING

A time series is a set of observations generated sequentially in time; the series may be continuous, or it may be discrete. From a stochastic point of view, streamflow (X_t) can be regarded as consisting of four components – namely trend (periodic, T_t or seasonal, S_t), correlation (K_t), and random (ϵ_t) components, which can be combined as follows (McMahon and Mein, 1978, 1986; Kottegoda, 1970):

$$X_t = T_t + S_t + K_t + \epsilon_t \tag{5.1}$$

To obtain representative stochastic data, it is necessary to identify and measure the strength of each component. A fifth probable component to the above model would relate to catastrophic events, such as extreme floods or droughts. To account for this additional component, a 'catastrophy factor' term – represented by δ – may be included in the above model in order to arrive at a more general form of the equation, viz.:

$$X_t = T_t + S_t + K_t + \epsilon_t + \delta \tag{5.2}$$

One characteristic of a time series relating the sequencing of the data is *persistence*; the common quantitative measure of persistence is serial correlation. In streamflow, persistence arises from natural catchment storage effects that tend to delay the runoff; thus, over a short time period, high flows in one interval will tend to be followed by high flows in the following interval. The longer the time period, the lesser the effect of persistence – and for many streams it is negligible for annual flows. As might be expected, the size of the river catchment will have a marked effect on the significance of this persistence – a fact that is even more appreciable when it is recognized that larger reservoir capacities are necessary on catchments with strong persistence in its runoff. This property is indeed very important in storage-yield analyses.

Another important concept of a time series is the notion of *stationarity*. A time series is considered stationary if the statistical properties characterizing it are time-invariant. Hydrologic time series are rarely stationary because of seasonal fluctuations in the mean and variance. However, it is frequently possible to isolate and remove the non-stationary aspects, leaving a stationary time series that can be modeled more efficiently. For instance, to partially eliminate the non-stationary effects of seasonality, monthly data can be standardized by use of the following type of simple transformation:

$$z_{jt} = \frac{x_{jt} - x_j}{s_j} \tag{5.3}$$

where z_{jt} = standardized monthly flows; x_{jt} = monthly flows; x_j = mean monthly flow for the j-th month; and s_j = standard deviation of monthly flows for the j-th month.

Detailed elaboration of the systematic approach to hydrologic time series modeling are provided elsewhere in the literature (e.g., Salas *et al.*, 1980). In general, any modeling of hydrologic time series involves making a decision on whether the model is to be univariate or multivariate; a combination of univariate and disaggregation models; combination of multivariate and disaggregation models; etc. For instance, in the simulation analysis of a reservoir, monthly inflows to such reservoir must be generated. If there is no other cooperating reservoir in the system acting jointly with the primary reservoir, then univariate modeling of monthly streamflows at or near the site under study can be selected. On the other hand, if other cooperating reservoirs are considered that are acting jointly with the primary reservoir under study, then multivariate modeling of monthly streamflows at the various sites is the obvious choice; an alternative choice to the multivariate modeling of monthly streamflows would be to select the multivariate modeling of annual series and then to subsequently use a disaggregation model to obtain the corresponding monthly flows.

Time series modeling find two main uses in hydrology and water resources – namely, for the generation of synthetic hydrologic time series, and for the forecasting of future hydrologic sequences. The former use is generally required for reservoir sizing and the determination of the reliabilities of the reservoir capacities to meet the target drafts, and for other such similar applications; the latter use is generally needed for short-term planning of reservoir operation, for real-time operations of reservoir/river basin systems, and for other similar applications.

5.1.2. MODELING TECHNIQUES

Several modeling techniques can be envisaged in the world of systems analysis – with the following general classes being the most identifiable in many instances:

1. Analytical Optimization Models and Techniques;
2. Probabilistic Models and Techniques;
3. Statistical Techniques; and
4. Simulation and Search or Sampling Techniques.

Although these categories are not disjoint, nor do they provide a complete coverage of the set of all systems analysis techniques, they may be considered as an adequate and broad-based classification for the purpose of hydrologic simulation studies.

Analytical Optimization Models and Techniques include the optimization methods that utilize classical calculus and Langrangian multipliers as well as mathematical programming and control theory. They are both descriptive and prescriptive.

Probabilistic Models and Techniques are a class of techniques that include and build on the elementary technique for describing stochastic system elements with appropriate parameters – for example, stochastic processes and techniques based on queuing and inventory theories. Probabilistic methods are used in the design of hydraulic structures in order to allow for the inclusion of a measure of the frequency of occurrence for the extreme events of high and low flows. With knowledge of the frequency of an event, the risk involved in specifying certain design standards for a project can be assessed.

Statistical Techniques may include such methodologies as Multivariate Analyses (via component analysis, regression analysis, factor analysis, discriminatory analysis, cluster analysis, canonical analysis, etc.), Statistical Inference, and Decision Theory. As a set of statistical tools that provide efficient methods for describing statistical interdependence between system/hydrological variables, multivariate techniques are often employed to define relationships between variables of a water resource system.

Simulation is a process for digesting data and evaluating production functions. A simulation may be time-sequenced, in which the analysis examines the state of the system at successive time intervals Δt, $2(\Delta t)$, $3(\Delta t)$,......, $n(\Delta t)$; or it may be event-sequenced, which takes note of the times whenever an event of specific interest occurs. A simulation may be *deterministic* or *stochastic* depending on the significance of randomness among input events, outputs, or relationships that specify the technological function.

In general, water resources planning problems are often complicated by multi-purpose multi-objective factors – and each of these characteristics has its effect on the nature of the systems study. Consequently, in most cases, there is no unique mathematical model for any particular problem; rather, a set of linked models is often more useful. The overall model may be nested in such a fashion that outputs of one model are inputs of another; there may also be feedback links where, for instance, 'Model#2' receives inputs from 'Model#1', and in turn the outputs of 'Model#2' are inputs to 'Model#1'. Indeed, models and their links configuration are usually problem-specific. Thus, care must be taken in coupling these models so that important relationships are not lost in the process. Notwithstanding the problem-specificity and unique attributes, a number of algorithms and associated computer codes, applicable to certain types of mathematical models and unique problem scenarios, are available as commercial 'standard packages' that can be adapted for case-specific applications.

5.2. Design of Reservoir Storage-Capacity with Inadequate Hydrologic Data

While the need for many types of hydrological and meteorological data for the design of water resources projects is obvious, it is often realized (especially in many developing regions) that such data are either lacking or inadequate. In the absence of very long historical records, sequential generation of hydrological data has been a tool for the hydrologist in creating synthetic records that could be used in planning efforts. But since the generated set of data is only as good as the historical set employed in such a synthesis, the historical set should not be too short. If the data is non-existent at all, then regional analyses procedures should be employed.

In general, inadequate data make it more likely that wrong decisions will be made, that wrong criteria will be selected, and that inappropriate and uneconomic designs will be adopted. The end product can be a system that partially or entirely fails to meet the anticipated objectives – with the benefits it produces bearing little positive relation to the capital invested. Consequently, where inadequate hydrological data is all that is available, an important consideration in the design task is the amount of risk that is perceived as being acceptable in selecting a reservoir size somewhat too large or too small for its streamflow supply – and therefore over- or under-estimating project accomplishments for this reason.

Indeed, hydrologic data of sufficiently long duration is generally essential for the selection of better decision criteria in design, planning, and operation of water resource projects that results in optimum development and that avoids the risks from system under-design or the uneconomic costs of system over-design. Typically, the worthiness of the data used should be assessed through a sensitivity analysis that considers the entire project and all the uncertainties and risks involved; such an analysis requires the utilization of hydraulic and hydrologic principles, systems analysis, and probability theory. Alternatively, the evaluation of the worth of data can be made by the use of Kalman filtering techniques (e.g., Chiu, 1978); Kalman filters provide a mechanism that integrates the mathematical and observation system models, and the information content of each new piece of observed data in making new estimates or prediction of system parameters and output.

Another important decision-making tool relates to the use of fuzzy information in water resources allocation problems. The concept of fuzzy set is an attempt to mathematically formalize information of imprecise nature to allow for its usability in modeling the real systems. The use of fuzzy set theory is particularly potent in the treatment of imprecision generated by the qualitative attributes of water resource systems.

5.2.1. THE NEED FOR SYNTHETIC DATA: STOCHASTIC GENERATION OF SYNTHETIC DATA AND FLOW GENERATION STRATEGIES

The use of generated data gives the designer a clearer picture of the possible variability of any estimate, and allows the analyst to choose a value that relates to the risk level that is acceptable to the designer. While it is not possible to create information that is not already in the record, it is possible to use the information more systematically and more effectively – because the system design can now be evaluated in a more statistically sophisticated manner.

In general, the complexity of natural streamflow is impossible to model. Instead, only a conceptually adequate model is adopted that is capable of reproducing certain flow properties and watershed characteristics that are deemed relevant to the problem on hand. In any case, the type of model(s) to use may be selected from quite a wide range of listings – including AR (autoregressive), ARMA (autoregressive-moving-average), ARIMA (autoregressive-integrated-moving-average), FGN (Fractional Gaussian Noise), BL (Broken Line), SL (Shifting Level), etc. models. Details and applicability of these and indeed several other related models can be found elsewhere in the hydrological literature (e.g., Salas *et al.*, 1980; Box and Jenkins, 1976; Clarke, 1973; Mandelbrot and Wallis, 1969; Jenkins and Watts, 1968; Kendall and Stuart, 1968; Harms and Campbell, 1967; Hammersley and Handscomb, 1965).

Model choice and parameter estimation are the keys to the effective use of stochastic streamflow sequences in the hydrologic design process. For short-term data the performance of models such as the *Thomas-Fiering Seasonal Model* (see, e.g., McMahon and Mein, 1978; Clarke, 1973; Fiering and Jackson, 1971; Fiering *et al.*, 1967) as a means of generating synthetic sequences, is adequate. Methods used for the synthesis of monthly flows vary with the nature of the non-stationarity observed – but the Thomas-Fiering model performs very well in this regard. With respect to non-seasonal schemes, one of the simplest generating models for streamflow synthesis is the annual *Markov Model*; the theoretical justification of the form of the algorithm for the non-seasonal Markov model, as well as some practical considerations to its application are elaborated elsewhere in the literature (e.g., McMahon and Mein, 1978). Strategies for the generation of shorter time intervals (e.g., daily, hourly, etc.) that might be needed in flood studies are also elaborated in greater detail elsewhere in the literature (e.g., Salas *et al.*, 1980 and Clarke, 1973).

If streamflows are stationary, but the available record is short, then it can sometimes be extended by use of methods based on correlation with flows at other sites at which longer records are available (Klemes, 1973; Matalas and Jacobs, 1964). Alternatively, regional estimates of streamflow statistics can often be developed that makes use of watershed characteristics (Thomas and Benson, 1970). Yet another approach that is applicable if land-use changes have modified streamflow characteristics is the use of rainfall-runoff models; stochastically generated precipitation may thus be converted to streamflow by use of an appropriate deterministic model.

Overall, synthetic sequences should be generated independently, each equal in length to the period of interest (corresponding to the expected useful/economic life of the system under study). One long sequence that is broken into sub-sequences yield interrelated sequences unsuitable for making probabilistic statements about expected system performance (Linsley *et al.*, 1982).

5.2.2 MULTIVARIATE STOCHASTIC MODELS

It is sometimes necessary to simultaneously generate sequences of several hydrological variables – e.g., of discharge at several gauging stations. Thus, it becomes necessary to construct multivariate models that preserves, not only the serial correlational structure for each variable, but also the cross-correlational structure between all pairs of variables. For instance, simulations of complex multireservoir systems require the generation of concurrent flows at many sites. Further, it is often necessary to consider the variations of flows within each year. Some of the multivariate (viz., multi-season multi-site) models that have found wide practical application are annotated below.

1. *The Disaggregation Model* (otherwise referred to as the *Method of Fragments*) – allows for the generation of synthetic flows that reproduce statistics both at the annual level and at the seasonal (namely, quarterly, monthly, weekly, or daily) levels. The method breaks down the input annual values that maintain all relevant statistics and preserve the logical algebraic relation with the annual statistics. Further elaboration on this modeling technique is provided in Chapter 8, and an even more detailed discussion can be found elsewhere in the literature (e.g., Linsley *et al.,* 1982; Stedinger, 1981; Salas *et al.,* 1980; Svanidze, 1980; Lane, 1979; Hoshi *et al.*, 1978; Valencia and Schaake, 1973). In fact, the disaggregation methodology is very flexible, and can be adapted to meet the needs of many situations encountered in water resource simulation studies. When flows at many sites or in many seasons are required, the size of the disaggregation model can be reduced by disaggregation of the flows in stages – and not attempting to explicitly reproduce every season-to-season correlation (Lane, 1979).

2. *The Multivariate Matalas Model* – which has been introduced by Matalas (1967) and further developed by Young (1968), is in fact the multivariate extension of the univariate first-order autoregressive, AR(l), model. Clarke (1973) provides a further generalization to form a multivariate second-order autoregression. Further elaboration of the first-order multivariate autoregressive model is considered in Chapter 8. An analogous multivariate Moving Average model is discussed in greater detail in Clarke (1973), and a multivariate extension to the Fractional Gaussian Noise model is elaborated in Matalas and Wallis (1971).

3. *Extension of Annual Flow Models* – comprises converting the time series of seasonal flows into a homogeneous sequence of normally-distributed zero-mean unit-variance random variables. These can then be modeled by an extension of the annual flow generators (Clarke, 1973).

Detailed discussion of the wide spectrum of multivariate generation strategies can be found elsewhere in the hydrological literature – including Linsley *et al.* (1982), Salas *et al.* (1980), Svanidze (1980), Lane (1979), Clarke (1973), Yevjevich (1972b).

5.3. Evolution and Philosophy of Stochastic Simulation in Reservoir Systems Modeling and Design

Simulation – regarded both as an art and as a science – is the solution of a management model by trial and error. As with optimization models, simulation models may be either deterministic or stochastic – one of the most useful tools in water resource systems planning being *stochastic simulation.* Stochastic simulation preserves the random nature of the system inputs (or inflows) – in addition to preserving the physical sequence of the process, in which one sees the evolution of the process through sequential computations.

Included in any stochastic simulation model is some provision for the generation of sequences of random numbers that represent particular values of events such as streamflows, rainfall, flood, etc. To generate a sequence of values for a random variable, probability distributions for the variables must be specified; historical data and an understanding of the physical processes are used to select appropriate distributions and to estimate their relevant parameters.

The random/stochastic nature of inflows to a reservoir system is treated as such by incorporating a stochastic inflow model – via an implicit stochastic approach, or by an explicit stochastic approach – directly within the simulation procedure. In the implicit approach, a number of synthetic inflow sequences are generated using a time series model, the system is optimized for each streamflow sequence, and operating rules are determined by applying multiple regression analysis to the optimized data sequences (Maidment and Chow, 1981). The explicit stochastic approach considers the probability distribution of the inflows rather than specific flow sequences; this approach facilitates the analysis of risk of failure of the system through the use of chance constraints or penalty functions to limit the risk of failure to an acceptable level (Sniedovich, 1979).

The general dynamic behavior of a system in discrete times can be expressed by the following vector difference equation:

$$X_{t+1} = AX_t + BU_t + GW_t \qquad t = 1,2,........T \tag{5.4}$$

where X_{t+1}, X_t are *n* vectors of state variables describing the system at beginning of time intervals $(t+1)$ and *t*, respectively; U_t is a *p* vector of input or control variables; W_t is an *r* vector of independent normal random variables, N(0,1); A, B, G are matrices of coefficients of dimension (*n* x *n*), (*n* x *p*), and (*n* x *r*) respectively; and *t* is a time or stage index.

The key point is that, the variation in the state of the system can be directly expressed as a function of the variation in the input variables. Indeed, this model fulfills the conditions for a Markov process – so that the state of the system in the next time interval depends on the current state of the system, and not on any states observed further back in time. This is demonstrable from the mean of the output state variable, X_{t+1}, given the values of the input state variable, X_t, as follows:

$$E\left[\frac{X_{t+1}}{X_t}\right] = AX_t + BU_t + G\{E[W_t]\} \tag{5.5}$$

Some authors (e.g., Maidment and Chow, 1981) have considered the optimization of the performance of such a system, with the objective to maximize the expected return or net benefits. For illustrative purpose, consider the monthly operation of a reservoir system in which S_t, S_{t+1} are the volumes of storage in the reservoir at the beginning of periods *t* and $(t+l)$ respectively; α is a coefficient accounting for seepage and evaporation losses; U_t is the volume of water released through the controlled outlet works in period *t*; and q_t is the volume of inflow in time period *t*. Then, ignoring uncontrolled flows over spillways, the continuity equation yields the following relationship:

$$S_{t+1} = \alpha S_t - U_t + q_t \tag{5.6}$$

If q_t is an independent and normally-distributed random variable with mean μ_t and standard deviation σ_t, this continuity equation may be rewritten in the general format of the vector difference equation as follows:

$$S_{t+1} = \alpha S_t + [-1,1]\begin{bmatrix} U_t \\ \mu_t \end{bmatrix} + \sigma_t w_t \tag{5.7}$$

For the independent inflow model:

$$E[S_{t+1}] = \alpha S_t - U_t + \mu_t \tag{5.8}$$

$$\mathrm{Var}[S_{t+1}] = \sigma_t^2 \tag{5.9}$$

It should be recognized that a simulation of reservoir performance on historic data alone gives only a single realization of what may happen in future situations. Consequently, multiple simulations from a number of independently generated flow series that have the same statistical properties of the parent historical data, but differ from the parent with respect to their detailed variation, is usually adopted. The outcome of such multiple simulations can indeed be expected to be more representative of the future performance of a reservoir than that given by the historical one alone. Furthermore, the variance of the storage yield estimate can be better appreciated from multiple simulations (O'Connel and Cole, 1980).

Simulation has indeed long been regarded as an important design aid to hydraulic engineers. In a well-designed simulation program, the designer can consider nonlinearities of system response; unique characteristics of economic decisions; uncertainties surrounding the hydrologic input and economic demands; long-term trends and cyclic variations among the inputs and outputs; a wide range of operating policies for the system; and indeed all manner of hydraulic and economic issues which are so difficult to incorporate into analytical formulations. It should be borne in mind, however, that this method of solution does not guarantee an optimal solution to the problem; it merely describes the outcome of a specified configuration. Also, it may become an expensive procedure if several many reservoir configurations are involved in the analysis.

5.4. Deterministic-Stochastic Hybrid Models

In a typical modeling exercise or procedure, the problem is not just to choose between deterministic and stochastic/probabilistic approaches. Rather, the issue is one of determining the conditions or circumstances that allows for the use of either approach, or even a combination of both. In fact, the distinction between deterministic and stochastic processes is only an artificial one – depending basically on the time and space scales used in the modeling effort. For instance, V. Nagy (1977) makes a differentiation between deterministic and 'indeterministic' models by relating them to the concepts of certainty and uncertainty; Yevjevich (1972b) considers the change from a stochastic process to a deterministic one that is based on the transitioning from a 'microscale' to a 'macroscale' in space and time.

In general, a random input to a deterministic system to give a random output can hardly be realistically described in terms of either stochastic or deterministic modeling – but this is better described by a marriage of the two into a hybrid model. In fact, stochastic methods in combination with deterministic methods offer the prospect of improving the overall system design process, and also for producing better outcomes.

5.5. Risk and Uncertainty in Reservoir Design

The *risk* associated a hydraulic component, subsystem, or a system is defined as the probability that the loading on this element exceeds the resistance – i.e., the probability of failure. Conversely, the *reliability* is the probability of survival – i.e., the resistance is greater than the loading.

Due to the stochastic nature of the inflow process, every combination of specified values of reservoir storage capacity and draft has associated with it a certain probability of an empty reservoir – giving rise to a situation where the reservoir is unable to fulfill all its demand; this probability is often termed the *risk of failure*. The complementary value to the risk of failure – the *reservoir performance reliability*, PR – is given by:

$$PR = 1 - (\text{Risk of Failure}) \qquad (5.10)$$

Typically, a reliability level and/or a risk level should be specified in a reservoir design. The reservoir is then designed to meet the pre-determined level, i.e., if project objectives are to be satisfied.

The problem of uncertainty and/or risk in reservoir management caused by the stochastic nature of streamflow have been investigated and reported by various researchers (e.g., ReVelle, 1999; Wurbs, 1996; Vogel and Bolognese, 1995; Marino and Mohammadi, 1983; Simonovic and Marino, 1982, 1981, 1980; Houck *et al.,* 1980; Hirsch, 1979; Houck, 1979; Hirsch *et al.,* 1977; Revelle *et al.,* 1969). In general, *uncertainty* is attributable to random measurement and computation errors of the available basic data; systematic errors; non-homogeneity in time; etc. Uncertainty tends to increase with increasing variance of sample data – and decreases as the sample length increases. The total of basic risk and uncertainty add up to give the overall hydrologic risk associated with the design project.

5.5.1. MODEL VERIFICATION AND PERFORMANCE

The accuracy of simulation depends on three factors – namely, the accuracy of input data, the effectiveness of parameter evaluation, and the inherent errors in the model (Linsley *et al.*, 1982). Before a data generation model is used in storage-yield analyses, it is necessary not only to check that it satisfactorily reproduces the main statistical characteristics defining the streamflow process, but also to check that critical periods are being satisfactorily generated. A general validation schedule to the duality in the analysis of a hydrologic time series is shown schematically in Figure 5.2. It is necessary to check how representative the generated records are in relation to the actual flows, as mirrored by the historical record. If traces generated are found to bear close parameter-resemblance to those of the natural flow record, then an adequate – though by no means complete – scheme may be indicative. In effect, this constitutes an assessment of the model performance. Some validation testing procedures that may be applied to annual or monthly models are elaborated in McMahon and Mein (1978).

Several model verification methods are based on goodness-of-fit criteria. Statistical *goodness-of-fit* constitute an objective and precise way of measuring the discrepancies between the model and the sample; the Chi-square and the Kolmogorov-Smirnov statistics are the most widely used of such tests (see, e.g., Salas *et al.*, 1980; Kite, 1977, Yevjevich, 1972a). Alternatively, distribution-free or nonparametric tests are often as powerful – maybe even more so than standard tests which assume that the observations are normally distributed; this is measured by the statistical significance of various results and hypotheses testing.

In general, when a mathematical model is used to represent a real-world system, and then the same input is applied to both the real system and the mathematical model, two different responses can be anticipated. The closeness of these responses, however, gives an indication of the goodness-of-fit and validity of the mathematical model.

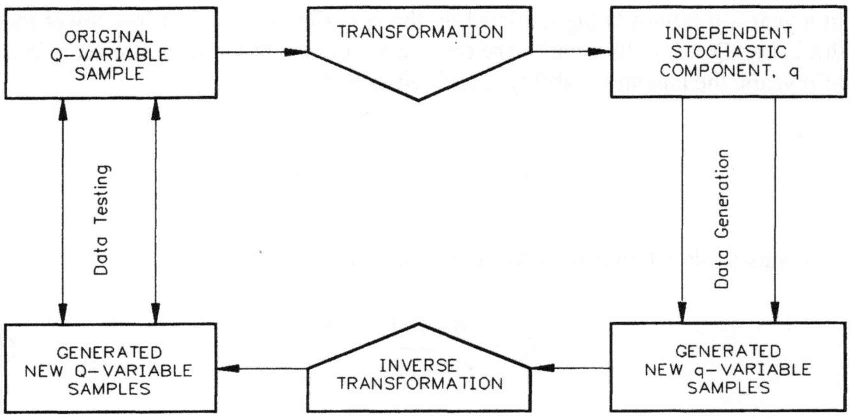

Figure 5.2. Duality in the analysis of hydrologic time series

5.5.2. RESERVOIR PERFORMANCE RELIABILITY

A water resources project is considered satisfactory if its performance achieves or exceeds a design expectation. On the other hand, the project is said to be unsatisfactory if the loading, L, (e.g., design floods) exceeds the resistance, R, of the project facilities (e.g., spillway and/or flood storage capacity of a dam/reservoir). Quite often, a single resistance can respond to several loadings; the converse is also true, as well as other complex interactions. The *probability of failure* (or the unsatisfactory performance probability) can then be defined by the following relationship:

$$Risk = P_{fail} = \Pr\{L > R\} = \int_{[(R,L)/(R<L)]} f_{R,L}(r,l)\,dr\,dl \tag{5.11}$$

This can be evaluated directly if the joint probability density function of R and L, $f_{R,L}(r,l)$, is known or is prescribed. P_{fail} is also equivalently evaluated as Pr{(R/L) < 1}, or Pr{(R-L) < 0}. In reality, R and L usually are functions of many component random parameters whose contributions in assessing the probability of successful performance of the project can be evaluated through an analysis of variance or the coefficient of variation; in a typical example of the application of the above formulation, Tang and Yen (1972) presents a first-order second moment reliability analysis of the performance of reservoirs.

The *reliability* of a reservoir is defined as the probability that, for a reservoir of specified size that is built, it will reasonably deliver the expected demand through its design/economic lifetime without failure. The reliability characteristics annotated below may be differentiated as part of a systematic approach to evaluate reservoir performance reliability (O'Connel and Cole, 1980; Klemes, 1973).

1. *Annual Reliability, R_a,* is the probability that no failure in water supply will occur in a year – a failure being depicted by the occurrence of an outflow lower than the draft (i.e., Q < D). In a stationary outflow process, and for *m* failure years in a total of *n* years, the annual reliability is defined as follows:

$$R_a = \frac{(n-m)}{n} \tag{5.12}$$

Sometimes this relationship takes the variant form of:

$$R_a = \frac{(n-m)}{(n-1)} \tag{5.13}$$

R_a is also referred to as *occurrence-based reliability* or *certainty*, and it represents the percentage of years of no-deficit occurrences.

2. *Time Reliability, R_t,* also known as *duration-based reliability*, is defined as operation time during which the demand was fully satisfied (i.e., no deficit, Q = D). It is the sum of all *non-failure periods*, Δt, as a fraction of the total operation period, T – and is represented as follows:

$$R_t = \frac{1}{T} \times \sum_{no\ deficit} \Delta t \tag{5.14}$$

3. *Volumetric Reliability, R_v,* is the actual delivered portion of the total volume of demand during a period T. It is also called the *quantity-based reliability* – and is represented by the following relationship:

$$R_v = \frac{Actual\ Supply}{Demand} = 1 - \frac{\int\limits_{Q<D} (D-Q)dt}{\int\limits_0^T DdT} = 1 - \frac{\sum \Delta V}{T \times D} \tag{5.15}$$

where ΔV is the quantity of short-fall within a period T for which the supply, Q, is curtailed below the constant draw-off rate, D.

In the case of a constant (and continuous in time) target draft, the following relationship holds true:

$$R_a \leq R_t \leq R_v \tag{5.16}$$

For more complex reservoir release rules, this relationship may not hold true. In general, storage-reliability depends on the operating period, the statistical properties of the streamflow, and the withdrawal required. The reliability with which a reservoir can sustain a specified draft (i.e., yield reliability) is indeed a problem of fundamental interest to the water resources systems analyst.

5.5.3. UNCERTAINTY ASSESSMENT VIA SENSITIVITY ANALYSIS

As with many hydraulic structures, the uncertainties associated with dams/reservoir design can generally be classified into hydraulic-, hydrologic-, structural- and economics-related categories of uncertainties. *Hydrologic uncertainties*, which is particularly relevant to the coverage in this title, may be further put into the following three categories (e.g., Wood and Rodriguez-Iturbe, 1975; Tang and Yen, 1972):

(1) *Inherent Uncertainty* – due to the stochastic nature of the hydrologic process;
(2) *Model Uncertainty* – resulting from the limited amount of data usually available for assessing the true random mechanisms of the hydrologic process; and
(3) *Parameter Uncertainty* – due to an insufficient amount of the data that is used to estimate the parameters utilized in an assumed model.

In general, uncertainty enters into a design analysis whenever the consequences of a decision cannot be foretold with confidence – which is to say, almost always. In hydrologic design and operating decisions, the results of which are influenced by chance or unknown factors such as the whims of the weather, ignoring uncertainty so as to simplify problems is obviously untenable. Thus, the problem should be analyzed in such a manner that allows uncertainty to be explicitly taken into account. Typically, decisions under uncertainty may be based on one or the other of the following principles: the *maximum-returns* principle; the *minimax-risk* principle; or the *probabilistic approach.* Decision theory allows the decision-maker to consider together, the uncertainty of the modeled process, the quantifying of the decision outcomes, and the preferences of these outcomes. Furthermore, Bayesian analysis offers a probabilistic framework by which the uncertainty in any design variable and the knowledge about it can be incorporated into the analysis.

A simple method for assessing the effect of uncertainty on system performance is to vary the magnitude of the more uncertain parameters or variables (singularly, or in combination) and then to examine the corresponding results. Such sensitivity analyses provide a means of identifying those parameters or variables to which system performance is particularly sensitive. Whether or not these uncertainties should be reduced by additional study or data collection depends on the cost of the additional information needed to reduce the uncertainty vis-à-vis the possible increase in benefits that such reduction in uncertainty may provide.

Misspecification of a parameter may result in significant *loss of economic efficiency* (LEE), also called *regret*; the LEE is the amount by which the actual costs of implementing the non-optimal decision exceeds the costs that would result if the optimal design were selected. Also, there can be some loss of economic efficiency, when parameter errors result in decisions that are very different from the optimal one. Sensitivity analysis considers the possible cost of making other than the optimal decision. It does not, however, address the question of what decision should be made when the future values of parameters and other variables that will critically affect a system performance cannot be determined with the desired accuracy before appropriate decisions are made.

5.6. The Hydro-Economics of Reservoir Design

The design of a water-resource project depends on an intimate interplay of economics and engineering considerations. For instance, the determination of least-cost reservoir combinations can be obtained by use of an economic model that utilizes curves of equal capital costs (i.e., iso-cost curves); curves of equal flood-peak reduction; etc. (James and Lee, 1971; Kuiper, 1971). Indeed, the economics of a reservoir network analysis is a function not only of construction costs but also of the operating policy adopted. An efficient operating policy minimizes the losses incurred when the network fails to meet its objectives – and consequently maximizes the net return on the investment. For reservoir design, an efficient policy implies minimization of reservoir capacity to meet its objectives at a specified reliability level. The discharge or operating policy must account not only for the uncertainty in the streamflows, but also for conflicting objectives of the network.

The project evaluation of the time stream of benefits and costs is yet another important area that deserves close attention during the overall project economics analysis. In evaluating and comparing project components, it often becomes apparent that the flows of costs and benefits to be added at different time periods are not directly commensurable. This is because capital has a time value whose implications must be carefully traced out in each project evaluation. Fundamental to the evaluation of the value of different time streams, therefore, is the concept of the time value of money. Discounting formulas are necessary for comparison of different time streams of costs and benefits. The *discount rate* is the time value that is used to render comparable, inputs and outputs occurring in different time periods; it is the rate of money growth. In practice, comparison between time streams of benefits and costs are expressed in terms of present values or of equivalent annual cost; for the same discount rate and project life, both methods are numerically equivalent. In general, the overall desirability of any project is highly sensitive to the choice of discount rate and project life. Detailed discussion of this subject can be found in various textbooks of economics, and also in literature available for engineering project economics (see, e.g., James and Lee, 1971; Kuiper, 1971).

5.6.1. ECONOMIC CONCEPTS IN RESERVOIR PLANNING

The inputs to a water development project, invariably, consists of the naturally occurring water, land, construction materials and manpower, as well as a variety of machinery and equipment; the outputs consist of water at the times and places required for one or more of the several multipurpose uses derivable from a reservoir. The economist often expresses the relationship between the inputs to, and outputs from a system by means of a *production function*. The concept of production function is indeed basic in the economic evaluation of many a design undertaking.

Production is defined in economics as any activity intended to convert resources of given forms and locations into other resources of forms and locations deemed more useful for purposes of further production and/or consumption. Basically, a production function expresses the technologic limits on the outputs attainable from given inputs; it, therefore, summarizes the technological data of a problem and gives all the possible plans of design and operation that are feasible and efficient.

The relationship between the storage capacity of a reservoir and its yield – called the *storage-yield function* – is apparently one of the most important ingredients of the hydrological aspect of its production function. This relationship depends, in a very complicated fashion, on the inflow and outflow hydrology regimes. At one extreme, if the seasonal patterns of natural inflow and of useful outflow are the same, and if the region is humid enough not to require over-year storage, then full use of available water can be obtained with no reservoir capacity other than that needed to produce adequate head for moving the water and generating energy. At the other extreme, if inflow and outflow are completely out of phase, the useful output of a system is limited to its storage capacity. Naturally, intermediate cases are the prevalent ones in real world situations.

5.6.2. BENEFIT-COST AND COST-EFFECTIVENESS ANALYSES *VERSUS* OPTIMAL DESIGN FROM MARGINAL ANALYSIS

Benefit-cost analysis is a method of evaluating the worth of a project – by computing and examining the ratio (or difference) of the total monetary benefits derived from the project to the total monetary cost associated with the project. From the optimization point of view, the benefit-cost analysis may be placed into the following two basic groups:

i) Fixed Cost Method – in which the costs are fixed and the objective is to maximize the benefits; and
ii) Fixed Benefit Method – in which case the benefits are specified *a priori* and the task is to minimize the costs.

For a feasible water resources development, an economic decision should be based on high benefit-cost ratios, as well as low interest rates. The rationale for benefit-cost analysis is predicated on the two fundamental economic concepts – i.e., scarcity and substitution. The classical procedure of benefit-cost analysis seeks to measure the excess of benefits over costs of any proposed project. If all costs and benefits are accurately measured, maximizing the net benefits corresponds to the economist's objective of economic efficiency.

Cost-Effective Analysis, on the other hand, seeks to find significant differences in the costs or resource requirements among the available alternatives for effecting one or more goals while also examining the beneficial effects (Carpenter, 1970). The cost effectiveness method is the trade-off between effectiveness and cost of one program compared to the same trade-off of some other program or variants of the same program.

In general, the optimal plan is not necessarily the one corresponding to minimum average cost. The optimal design is the one for which marginal cost (i.e., the rate of change of cost with respect to output) equals marginal benefit (i.e., the contribution of gross benefits per additional unit of output). Marginal analysis procedures are indeed effective in testing combinations that are on the production function, and near the optimum. The method enables the analyst to adjust reservoir capacities and allocations in order to arrive at the best combination, and also to verify whether or not the target output levels are optimal. Unfortunately, this method can be slow and cumbersome, since only two variables are typically changed at a time.

Finally, it should be mentioned here that, invariably, cost functions in water resource systems do not show the marginally increasing costs on which economic theory is based – especially because of the following reasons:

i) Cost functions of various water resource systems components show economy of scale – i.e., the cost per unit capacity decreases as the capacity increases; and
ii) There is a finite discontinuity of the cost functions at zero capacity – with the investment costs being much larger than the Operations, Maintenance and Repairs (OMR) costs.

A more detailed discussion of benefit and cost estimation strategies, and indeed the analyses relevant to reservoir design problems can be found elsewhere in the literature (e.g., Loucks *et al.,* 1981; Howe, 1971; James and Lee, 1971; and Kuiper, 1971).

5.7. Optimization Techniques as a Design Tool for Water Resource Systems

In a classical sense, *optimization* can be defined as the art of obtaining the best policies to satisfy certain pre-set objectives – whiles at the same time satisfying certain fixed requirements or constraints. The process of selecting that set of decision or policy variables which 'extremizes' (i.e., maximizes or minimizes, as appropriate) the objective function (also called profit function, performance index, or return function) subject to the system's constraints is called the optimization procedure. Usually the constraints represent either system limitations, or the physical, technical, legal, and economic laws that the variables must satisfy.

5.7.1. THE OPTIMIZATION PROBLEM

Most optimization problems require the maximization or minimization of a given objective function subject to a set of constraints and a set of bounds on the policy variables. The general single-objective optimization problem with *n* decision variables and *m* constraints may be stated as follows:

$$\text{Extremize} \quad z(x_1, x_2, \ldots\ldots, x_n)$$

$$\text{subject to} \quad g_i(x_1, x_2 \ldots\ldots, x_n) \leq 0, \quad i = 1,2,\ldots m$$
$$x_j \geq 0, \quad j = 1,2,\ldots n \qquad (5.17)$$

Another typical problem in water resources planning studies involves minimizing the maximum of a series of unknown decision variables.

The utility of a given optimization technique may depend strongly on the dimensionality of the problem. High dimensionality tends pose impracticality problems that gives rise to the notion of the 'curse of dimensionality.' Due to the dimensionality problem that often becomes a significant issue, 'decomposition' usually becomes essential in the application of the various techniques to a multiple reservoir system.

Overall, extensive literature review typically reveals that no general and/or unique algorithm exists for the solution of reservoir optimization problems. The choice of method usually depends on the characteristics of the system, the availability of data, and the objectives and constraints specified for the problem.

5.7.2. SOLUTION TECHNIQUES FOR THE OPTIMIZATION PROBLEM

Optimization problems may be solved via one of several solution techniques – such as, Calculus; Linear Programming (LP); Non-linear Programming (NLP) (e.g., Direct Search, Lagrange Multipliers, Penalty Functions, Gradient Search Methods, Quadratic Programming, Geometric Programming, etc.); Optimal Control Theory (OCT); Penalty Successive LP (PSLP); Network Flow Programming; Dynamic Programming (DP); Simulation (which, strictly speaking, is not an optimizing procedure – because simulation models do not necessarily identify optimal designs); Decomposition and Multilevel Approach; etc. A comprehensive discussion of these techniques and their extensions are provided elsewhere in the literature (e.g., Mousari and Ramamurthy, 2000; ReVelle, 1999; Mays, 1996; Wurbs, 1996; Johnson *et al.* 1991; Yakowitz, 1982; Loucks *et al.,* 1981; Haimes, 1977; Haimes *et al.,* 1975; Gottfried and Weisman, 1974; Morin, 1973; Wismer, 1971; Lasdon, 1970; Bellman, 1957). Some of the primary methods that have found wide applications are briefly elaborated below.

- *Linear Programming.* LP problems include the optimization of a linear, algebraic objective function, subject to a number of linear, algebraic constraints. LP is a very effective optimization technique for solving linear static models. LP problems require that the independent decision variables are non-negative – otherwise they will vary continuously, making the problem meaningless. Unlike most other optimization techniques, LP packages are widely available at most of the major scientific facilities throughout the world. To use LP, it is not necessary to understand all the details of the LP solution procedures – a distinct advantage over most other types of optimization.

- *Non-linear Programming.* A nonlinear programming formulation has distinct advantages over a LP formulation in view of the fact that it enables the designer/analyst to consider the economies of scale. The Kuhn-Tucker conditions provide much of the basis/foundations for nonlinear programming (NLP) (e.g., Wismer, 1971; Lasdon, 1970).

- *Dynamic Programming.* DP has been defined as the theory of multistage decision processes (Bellman, 1957). It is about the most used nonlinear optimization technique in water resource systems. DP is seen to possess certain important advantages for the analysis of water resources systems – especially because it can treat the non-convex, nonlinear, discontinuous objective and constraint functions with no difficulty. Furthermore, DP methods are generally more amenable to stochastic inputs than are the other procedures – and it is also ideally suited to system decomposition. DP is indeed quite a flexible method, and solutions do not depend on any particular algorithm. However, the 'curse of dimensionality' is a major shortcoming of DP – albeit to some extent, some variations of the DP overcome this problem. Some of the well-studied variations of the conventional DP include the Discrete DP (DDP), the State Increment DP, Discrete Differential DP (DDDP) and Differential DP. Yakowitz (1982) presents a very comprehensive survey of DP models for water resource problems, as well as examine some of the computational techniques involved.

In reality, there are only a few problems that are entirely linear. For instance, the problem of flood control cannot be treated judiciously in a LP formulation. However, the availability of powerful LP procedures and tools provides a major incentive for approximating problems in this form. Hence, one of the better means for solving NLP problems is replacement of the nonlinear relationships by linear approximations – called linearization – and then subsequently solving the resulting LP problem. When the nonlinearities are small, the result may be a very close approximation to the true solution; however, it becomes inadequate as the nonlinearities become significant.

Overall, many realistic and practical optimization problems tend to be large with significant number of variables and constraints. When this occurs, one naturally considers how the complex optimization problem can be reduced to a series of less complex sub-problems that can be optimized separately, and then sequentially coupled to one another; the success of such an effort depends on the mathematical structure of the system being optimized. If one or another of the sub-problems has no feasible solution, then the problem as a whole has no feasible solution. Furthermore, in the event of an unbounded i-th sub-problem, an artificial bound may be applied.

Although water resources systems must be analyzed for optimality insofar as possible by considering a complete system and its objectives, a decomposition into subsystems will both simplify the study of complex systems, and provide feasible methodologies for analysis. Generally, the decomposition is accomplished by introducing new variables – called pseudo-variables (i.e., 'false' variables) into the system. Subsequently, each subsystem in the hierarchy is optimized separately and independently – applying different optimization techniques, as and when necessary; this depends on the nature of the subsystem models, as well as the objectives and constraints of the subsystems. Ultimately, the subsystems are joined together by coupling variables that are manipulated at a second- or higher-level in order to arrive at the optimal solution for the whole system. Indeed, decomposition and multilevel optimization have several significant advantages in the solution of large-scale complex optimization problems over conventional optimization methods; its attributes are discussed by Haimes (1977). In a hierarchical structure, different objective functions can be assigned each subsystem, leading to a multiobjective optimization analysis; when the objective functions are incommensurable, then the Surrogate Worth Trade-off (SWT) method can be successfully employed (Haimes *et al.*, 1975).

It is noteworthy that, even where optimization can be applied to only certain components of the system, such sub-optimizations can be of considerable value in improving the overall performance of the planning and design program. It is not always conclusive, however, that optimal analysis of all of the subsystems will produce an optimal solution for the total system; constraints, value trades, and certain incompatibilities will usually preclude such optimality from being achieved.

5.7.3. MULTIOBJECTIVE OPTIMIZATION IN RESERVOIR DESIGN

Multiobjective optimization has become prominent because most reservoirs are designed to serve more than one purpose. Multiobjective programing and planning is concerned with decision-making problems in which there are several conflicting objectives. It is noteworthy that, multiobjective analyses do not yield single optimal solutions, but are more useful in identifying the trade-offs among conflicting non-commensurable objectives.

The vector optimization model is a concise way of formulating a multi-objective problem. The general multiobjective optimization problem with *n* decision variables, *m* constraints, and *p* objectives is stated as follows:

$$\text{Extremize } \{Z(x_1, x_2, \ldots\ldots\ldots x_n) = [Z_1(x_1, x_2, \ldots\ldots\ldots x_n), Z_2(x_1, x_2, \ldots\ldots\ldots x_n), \ldots \ldots\ldots\ldots Z_p(x_1, x_2, \ldots\ldots\ldots x_n)]\}$$

$$\text{such that } g_i(x_1, x_2, \ldots\ldots\ldots x_n) \leq 0, \; i = 1,2,\ldots\ldots m$$
$$x_j \geq 0, \; j = 1,2,\ldots\ldots n \tag{5.18}$$

where $Z(x_1, x_2, \ldots\ldots x_n)$ is the multiobjective function, and $Z_1(\ldots\ldots), Z_2(\ldots\ldots), \ldots Z_p(\ldots\ldots)$ are the *p* individual objective functions. 'Extremizing' the problem implies that, as appropriate, either a minimization or a maximization is required. In this process, the amount of one objective that must be sacrificed in order to realize an increase in the other objective is what is called a *tradeoff*. The concept of tradeoffs is indeed of an inevitable importance in multiobjective programming problems.

5.7.4. THE NET-BENEFIT FUNCTION UNDER OPTIMALITY CONDITIONS

In defining the objective functions, denote the total net benefits that each interest group or user *j* derives from a project at each location, *i*, by $NB_{i,j}$. The $NB_{i,j}$ terms are defined by the benefits derived (if any) from the target allocation to each water use, T_i, less any losses and costs. Thence, where appropriate, the following relationship holds for all of the user groups, *j*, and locations, *i*:

$$NB_{i,j} = \sum_j B_{i,j}(T_i) - \sum_t \left[L_{i,j,t}(D_{i,t}) - G_{i,j,t}(E_{i,t})\right] - \sum_j C_{i,j}(K_i) \tag{5.19}$$

where $B_{i,j}(T_i)$ = net benefits to group *j* from an annual target allocation of T_i at site *i*; $L_{i,j,t}(D_{i,t})$ = losses to group *j* from an allocation deficit of $D_{i,t}$ at site *i* in period *t*; $G_{i,j,t}(E_{i,t})$ = gain (or loss, if negative) to group *j* from any excess allocation $E_{i,t}$ at site *i* in period *t*; and $C_{i,j}(K_i)$ = annual cost to group *j* of the capacity K_i of the reservoir project at site *i* (including capital and OMR costs). Generally, the distribution of $NB_{i,j}$ among the various users is influenced by the level of development of each project site, *i*. Alternative distributions of $NB_{i,j}$ can therefore be derived by the use of weighting factors, w_j, associated with each user *j*. The overall objective would then be to maximize the total weighted net benefits, viz.:

$$Maximize \sum_j w_j \sum_i NB_{i,j} \tag{5.20}$$

An alternative formulation would comprise of minimizing a loss function for the system.

5.8. Optimization Under Uncertainty and Risk

When we know what the various possible states of the world/system are, and have some means of estimating the probability of these states occurring, then we are faced with a decision under *risk*. If however, we know what the various possible states of the world/system are, but are unable to estimate the probability that a given state will occur, then we are faced with a decision under *uncertainty*. Uncertain situations may arise when there is no past information on which to base predictions, or when past information cannot be trusted to be a reliable guide. Risk-type situations are perhaps more common than uncertain ones.

Optimization under risk implies that one or more of the quantities affecting the objective function will be a random variable. It is possible to formulate and, under certain conditions, to solve optimization problems that contain random variables. This is generally accomplished by incorporating characteristics of the frequency functions (also called distribution functions or probability density functions) – such as the mean and standard deviation – into the mathematical model. Optimization problems of this type are called *stochastic programming* problems.

Stochastic Programming serves as an important analytical tool, especially because water resources systems are never strictly deterministic. When the objective function contains randomly varying quantities, we cannot assign a definite value to this objective function. In fact, the value of the objective function is itself a randomly varying quantity, which can be described by a frequency function $f(y)$. We can, however, seek to maximize (or minimize) the most probable (or expected) value of the objective function. If y represents the value of the objective function without consideration of risk, then upon consideration of risk, the objective becomes:

$$E(y)=\mu_y=\int_{-\infty}^{\infty} y\,f(y)dy \tag{5.21}$$

An estimate of the probable loss in the objective function based upon values less than the most probable value is given by:

$$\int_{-\infty}^{\mu_y}[\mu_y - y]f(y)dy \tag{5.22}$$

This is exactly counter-balanced by the possible benefit resulting from values of the objective function in excess of the most probable value, viz.:

$$\int_{\mu_y}^{\infty}[y-\mu_y]f(y)dy \tag{5.23}$$

Random variables may also appear in our constraints. When this occurs, we must also introduce the characteristics of the frequency functions describing these variables into the constraints (Gottschalk *et al.*, 1971; Tang and Yen, 1972; Yevjevich, 1972b).

In general, random variables are most commonly introduced into optimization problems through the objective function; the objective function usually is some economic index – that is determined by such issues as relates to prices, costs, and demands. Typically, such quantities can only be estimated, and often they must be regarded as having appreciable variability. The quantities in the constraint equations may, however be much better known, and often they can be considered as deterministic. Other sources of variability include: decision variables, technological coefficients, and constraint variability. When the variations in technological coefficients, decision variables, and constraint correlations are significant, consideration in a deterministic manner is no longer meaningful – calling for the use of probability constraints in what is called *chance-constrained programming.* Chance-constrained programming requires that probability limits be established for each constraint. However, no guidance is furnished to help establish these limits. Limits that are unduly restrictive may unnecessarily increase costs; limits that are too loose may lead to undesirable risks.

In general, the most useful procedure of choice for optimization under conditions of risk depends upon the particular problem; in all cases, the problem involves maximizing or minimizing the expected value of some quantity. Deterministic constraints have little meaning under conditions of risk. Whenever it is possible to assign a cost penalty to the violation of a constraint, it is desirable thing to do. The probable cost of violating this constraint can then be included in the objective function as a failure penalty. Use of the failure-penalty concept will generally avoid solutions that are unnecessarily conservative. In some situations, however, it is extremely difficult to assign a cost penalty when a constraint is violated. For sequential decision under risk, it is desirable to consider DP – since the initial DP solution provides the policy to be followed for all the possible events.

Chapter 6

HYDROLOGIC ANALYSIS OF FLOOD FLOWS

Floods and its devastating effects have been known from the beginning of life on earth. Its prevention or mitigation is therefore of high importance – to ensure a sustained human and ecological life. The purpose of flood control is to eliminate or reduce damage caused by the flooding of areas adjacent to rivers, or by the overtopping of engineering structures like dams, embankments or bridges. Flood control may be accomplished by means of the following general types of hydraulic structures:

1. *Reservoirs* – where part of the flood waters can be temporarily retained to help reduce the flood peak, and the passage of water masses is stretched over a longer period of time at a reduced flow rate; or
2. *Levees* – built along rivers to protect the adjoining areas by confining the flood to a particular channel, and in which case the flood peak may be increased because of the elimination of storage in areas that otherwise would have been inundated.

It is noteworthy that, in both of the above-indicated cases, the original travel time of a flood will be changed, and this may have serious consequences on downstream portions of the river. For instance, it could well happen that, by reducing flood peaks on a tributary, flood peaks on the main river downstream from the confluence are increased due to a change in the wave travel times – causing the floods on the tributary and the main stream to coincide.

6.1. Design Flood Determination

One of the most important tasks necessary for the design of a successful flood defense system, and for the effectual utilization of water resources, involves a determination of the critical or design flood flows and the elevation (i.e., height) to which the relevant hydraulic structures should be constructed. The *design flood* is that flow rate that a flood damage reduction project (such as a levee or a reservoir) is based upon – often greater than or equal to the so-called 100-year flood. In fact, the design flood is not only a physical term, but also has a public policy, political, legal, economic, and technical dimensions to it. Decisions made are directed at determining the extent to which the responsible policymakers wish to, or are able to, protect populations, the environment, and the wealth of society against the hazards of flood. Thus, the fundamental question in flood control programs may be stated simply as follows: 'what is the limit or threshold above which the area prone to flood will be put at risk, and what is the allowable risk level acceptable to society?' The design flood is indeed a very important hydrological parameter, especially because of its implication in determining both the safety of dams and in the economic viability of dams.

There is a very wide range of methods available for estimating a design flood for a given level of acceptable risk; the relevant level of risk usually is predicated on the different dam situations in terms of size and potential hazards. The most frequently used method for selecting the design flood is based on the probability of occurrence of the flood. From a hydrologic point of view, the basic question to be answered here is: 'what shall be the probability of occurrence of the flood against which the flood-prone area, the dam structures, and the downstream river sections will be protected at a pre-determined level of safety?' It is noteworthy that, once the basic appropriate design parameters have been determined, the designers and operators of flood control structures can only be made legally responsible for a flood-damage that occurs at floods lower than the accepted design flood (Mosonyi, 1999) – albeit it is not possible to assign an exact value to design flood because of its multidimensional attributes.

6.1.1. THE SELECTION OF AN ACCEPTABLE RISK LEVEL

The fundamental requirements in dam design for flood control may be stated as follows:

- there should be no loss of life anticipated;
- there should be minimal economic loss; and
- there should be no adverse environmental impacts.

The goal for ensuring 'no loss of life' may include a policy that there be no permanent structures for human habitation in the flood plain. In other situations, however, there may be large communities below a dam that is used for flood protection. Under such circumstances, as part of the dam safety evaluation, the *'probable maximum flood'* (defined below in Section 6.1.4) has to be considered for use as a design flood. In general, an incremental hazard category should typically be selected – which means at least a 200-year flood, or equivalently, one with an annual probability of occurrence of 1 in 200.

In fact, in contemporary engineering design practice, dams are expected to have a lifetime of 100 years or more – and consequently the flood probability is also selected on that basis. Thus, the 1 in 100 annual probability (or equivalently, the 100-year design flood) is an appropriate design parameter – i.e., if we expect a 100-year design lifetime. It is noteworthy that the use of years may be confusing to non-technical managers or stakeholders because they may wrongly assume that it is a flood that will only occur once in every 100 years – and appropriate clarification should be made to assist non-technical policymakers or other stakeholders in this regard.

6.1.2. THE CALCULATION OF THE RISK OF OVERTOPPING

The probability of exceedance of specific flood flows can be characterized by the return period, T, of the flood. The *return period* is the average time interval between occurrences of a hydrological event (such as a flood) of a given or greater magnitude, usually expressed in years. For example, a flood event with a probability of exceedance, *p*, of 5% (or 0.05) can occur in an average time period defined as follows:

$$T = \frac{1}{p} = \frac{1}{0.05} = \frac{1}{5\%} = \frac{100}{5} = 20\,years \qquad (6.1)$$

Ultimately, the risk of overtopping is a function not only of the return period of the flood, but also of the lifetime, N, for which the structure is exposed to that risk. The risk, R, of a flood of return period, T, being equaled or exceeded at least once within a time period, N, is defined by the following relationship:

$$R = 1 - (1 - \frac{1}{T})^N \qquad (6.2)$$

A schematic of the functional relationship for the risk of overtopping, i.e., R = *f*(T,N), is shown in Figure 6.1. In general, it may also become necessary to analyze the partial series data, together with the seasonal data, in order to come up with the appropriate design flood.

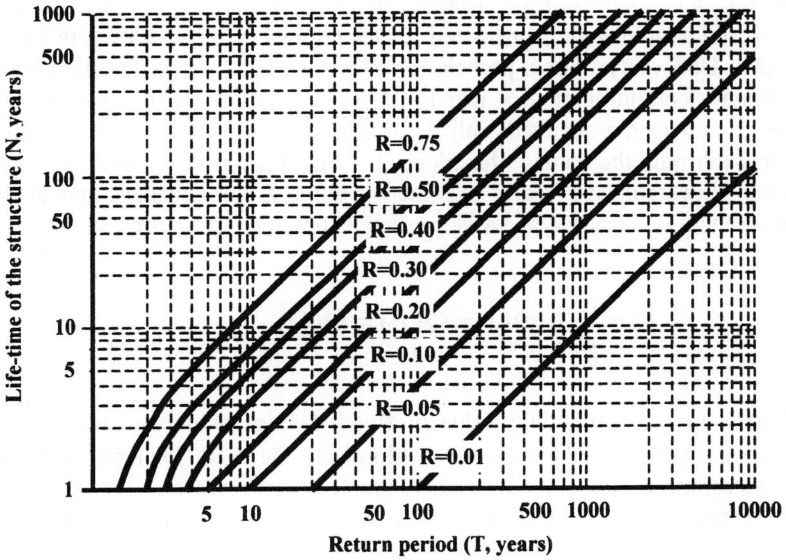

Figure 6.1. Risk of failure

Overall, an evaluation of the total risk of the hydraulic structure being overtopped involves integration of equation (6.2) over the full length of construction – taking into account specific details of the construction program, the return period of the design flood, and the seasonal nature of the floods. In this regard, a special issue that should not be overlooked relates to the hydrological risk associated with the duration of the construction period of the dam – which may be about 4 or 5 years, or even longer. In general, it would seem appropriate to adopt a lesser risk during the construction phase than for the post-construction era when the dam is in-place. Using the traditional approach, therefore, if for example, N = 100 years and T = 200 years, then in this example, the probability of occurrence of the critical flood is 0.4 (or, 40%). If the construction period is 5 years, then in the case of a risk of failure of 40%, the return period is T = 10 year. This means that, in the planning of structures to ensure safety during construction, floods having a T = 10-year return period should be taken into account in the pre-construction analyses.

6.1.3. HYDRO-ECONOMIC IMPACT ANALYSIS

After an evaluation of the risk of overtopping and a damage assessment, an economic analysis of the potential impacts of a given flood event usually is also necessary. One possible way to carry out this type of analysis requires the assumption of a continuously changing characteristic flood flow – i.e., starting with the discharge value, Q_o, and yielding flows of Q_M.

To illustrate the general process involved in the economic analysis, let us consider a given discharge, Q_M, and let us also assume that the hydraulic control structures have been dimensioned to handle flow volumes corresponding to Q_M. Consequently, if the maximum flow, $Q_{max} < Q_M$, then the structure and the downstream river section will be kept safe from the flood hazards. That is, this level of defense will prevent potential damages – and that can be considered as an equivalent or a surrogate benefit, B (expressed in US$/year). For a probability of non-exceedance, *p*, the discharges $Q < Q_M$ will result a certain amount of damage, D (expressed in US$). Thus, if we know the values of damages associated with the different levels of yearly floods, then it is possible to determine the damage function, D = *f*(p) (Figure 6.2) – and the integral of this function represents the yearly average benefit (or, the expected annual damage (EAD) avoidance) of using Q_M as basis for dimensioning the structures, viz.:

$$B = \int_0^p D(p)dp = \Phi(p) = \Phi(Q) \tag{6.3}$$

where D(p) is the corresponding value of flood damage associated with a flood of annual probability of exceedance of *p*. Alongside the Φ(Q) benefit function, it is generally necessary to also determine a cost function that represents the yearly costs of defense (expressed in USD/year) (i.e., relating to construction, operation, and maintenance) for a changing value of Q_M, viz.:

$$C = \Psi(Q) \tag{6.4}$$

The condition for arriving at the most economical discharge value is represented by the maximum of the difference between Ψ(Q) and Φ(Q), viz.:

$$\mathrm{Max}\{[\Psi(Q) - \Phi(Q)]\} \tag{6.5}$$

In general, however, when the damage is consistently and extremely higher than the benefits, then from a policy decisions point of view, a more favorable solution is the minimum of the [Ψ(Q))-Φ(Q)] value.

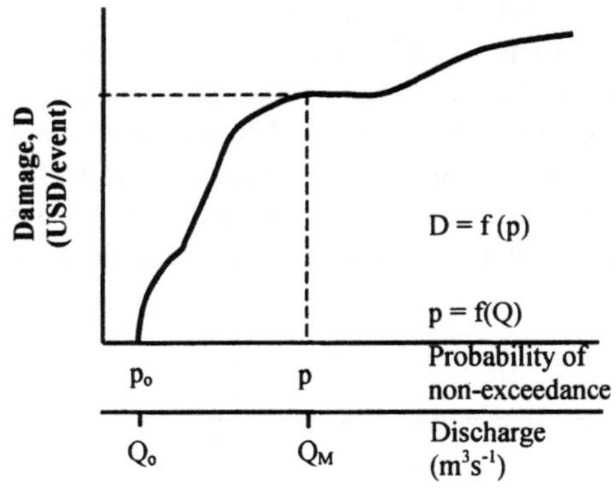

Figure 6.2. Economic damage evaluation (after Mosonyi, 1999)

Overall, if we choose an arbitrary value of Q_M for the design flood, then the value of yearly average damage can be represented by the sum of partial damages resulting from flows of magnitude Q_0 to Q_M. A procedure proposed by Mosonyi (1999) allows preliminary estimates of the expected annual damage via a series of experimental analyses. In the experiment, the distribution of annual maximum discharge is sampled randomly to generate an annual flood, and the annual damage is found via transformation using the discharge-damage functions (Griffin, 1996).

6.1.4. MODELING THE FLOOD FLOWS

The selection of the appropriate flood model is a basic question in the determination of a safe flood-carrying capacity for dams – often accomplished via flood modeling exercises. It is apparent that earlier types of methods for flood modeling have been based on a deterministic approach – where the basic concept rests on the methods of unit hydrograph analysis. This approach made it possible to estimate flood flows from rainfall storm events. Concurrently, a number of routing methods of various degrees of complexity were developed for rivers and reservoirs. In recent times, the concepts of *probable maximum precipitation* (PMP) and *probable maximum flood* (PMF), which define the expected physical upper limit to flooding, are widely used in the evaluation of the hazard of dams – albeit it often becomes very costly to utilize this method of approach, and also some significant sources of uncertainties remain with their determination. Furthermore, there are other different types of sophisticated models used for computing the flood hydrographs – with the objective being to describe the hydrologic cycle during flood periods. The applicability of such numerical models depends on the quantity and quality of basic data available in given river section.

More contemporary models are based on a probabilistic approach in which the systematic evaluation of stream gage records and maximum flood values allows for the use of the theoretical probability distribution functions (e.g., Pearson, Galton, Gumbel, Frechet, Krickij-Menkel, and Mosonyi) in the analysis of the extreme values associated with the flood distribution problem. Typically, such analysis results in a probability distribution function that generally does not have an upper bound (Figure 6.3). The methods may be applied to the peak flood discharge as well as the flood volume. The adequacy of the chosen probability distribution function may be determined by application of the appropriate 'goodness-of-fit' statistical tests – also including an analysis of uncertainty of the discharge-probability function. Indeed, the different distribution functions may yield different flood flows for the same river section and for the same period of time. Consequently, it is very important to explicitly indicate the type of distribution function used in any application, as well as the method of determination of the parameters – including the uncertainties associated with their estimations.

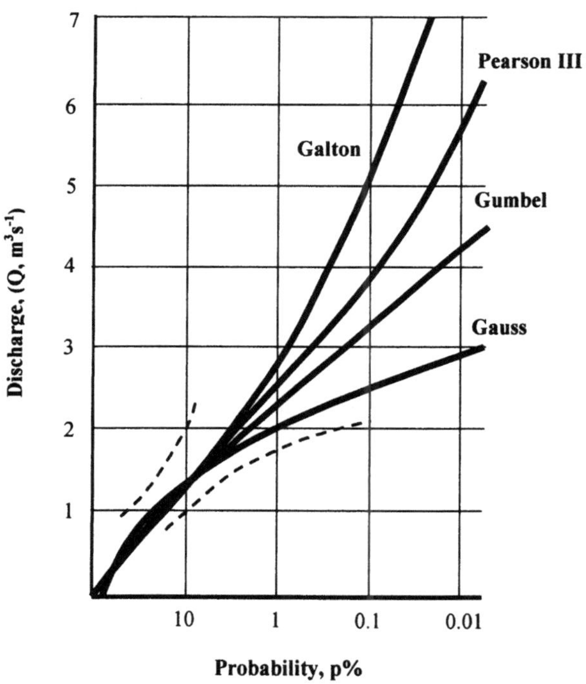

Figure 6.3. Comparison of the results of a flood flow analysis using different probability distribution functions

6.2. The Probability of Occurrence of Flood Flows

In flood flow studies, typically the investigated random variable in the given cross-section of the river is the maximum value of the annual flow, NQ, and the overarching objective in such flood studies is the determination of the probability distribution function of non-exceedance of this maximum flow, viz.:

$$F(X) = P(NQ \leq X) \tag{6.6}$$

The complementary function of exceedance is represented by the following equivalent relationship:

$$F^*(X) = P(NQ \geq X) = 1 - P(NQ \leq X) = 1 - F(X) \tag{6.7}$$

Typically, these estimates are made on the basis of the usually limited measured time series of discharges. In any case, crucial decisions rest on what the probability of occurrence of the flood shall be against which the dam and the flood-prone area will be protected, and also the determination the level of safety from this protection. To arrive at successful decisions, a theoretical probability distribution function that adequately fits to the measured data should be selected. The most frequently used method is based on the Pearson III-type of probability distribution – that requires a time series of at least 30 years. For such probability distribution function that generally does not have an upper bound, its use presents a problem when the results obtained by extrapolating a given sample are compared with the values obtained by deterministic methods. Indeed, the results obtained agree well with the measured data only up to the point when the probabilities corresponding to the length of the hydrographs are available. However, when the task is to determine the flood discharges of small probabilities (e.g., 0.01, 0.001, etc.), then the different methods tend to produce significantly different results in this range (refer back to Figure 6.3 above). Thus, the following considerations can be used in the flood analysis – that utilizes deterministic processes in the evaluation of the flood discharges of small probabilities:

1. Historical records often contain reasonable and conclusive information to use in the determination of flood conditions that can be expected for the project location.
2. The values of the possible maximum flood discharges can be estimated by analyzing the pertinent factors producing a characteristic flood wave. In the case of flood waves occurring in major streams, it is possible to obtain a good estimate by assuming the coincidence of the flood waves of the major tributaries observed so far.

Finally, based on all available information relevant to the evaluation of the maximum flood discharges, a theoretical probability distribution function that adequately fits the measured data can be selected for the further analysis of the flood flows.

6.2.1. ESTIMATION OF THE PROBABILITY DISTRIBUTION FUNCTION OF THE MAXIMUM FLOOD FLOWS FOR LARGE RIVERS

For a large river with tributary streams, the maximum flood flow is given by:

$$NQ = \sum_{i=1}^{n} Q_i \tag{6.8}$$

where Q_i are the discharges of the tributary streams. It is assumed that the conditional probability of each pair of data equals the probability of that data, viz.:

$$P(Q_i | Q_j) = P(Q_i) \tag{6.9}$$

This means that the spatial values of the measured discharges in the main river are independent random variables.

Application of the Gaussian Distribution Function

If the number of the tributary streams is $n \geq 15$, then the sum of discharges can be characterized by the normal probability distribution function, viz.:

$$F(x) = P(NQ \leq x) = N(m, \sigma) = \frac{1}{\sqrt{2\pi\sigma}} \cdot \int_{-\infty}^{x} e^{-\frac{(x-m)^2}{2\sigma^2}} dx \tag{6.10}$$

In this case, there is no need for a specific procedure in the determination of the parameters and functional values – because we simply can use the usual procedures for handling the normal probability distribution function.

Practical experience relating to the test of normality, and also the numerical and graphical fitting of data to the Gaussian system of coordinates have come to show that, in the case of catchments areas of $A \geq 100{,}000$ km^2, the flood flow of the main river carrying the cumulative discharges of the tributaries follows the normal probability distribution function. The normality is typically verified by using the single Kolmogorov test, the χ^2 test, or the Gauss method. Without reproducing the complete mechanics of the calculation processes, an illustrative example that used the Kolmogorov-Smirnov test to validate the applicability of the normal distribution to the maximum annual discharge records (1901 to 1992) for the Mohacs River section of the Danube River is shown in Figure 6.4.

Application of the Beta Distribution Function

An idea from Mosonyi (1999) for the application of Beta probability distribution function is presented in Figure 6.5. Typically, the cumulative distribution function is a monotonous non-decreasing function of *x*. However, the random variables (i.e., discharges) used in practice are usually bounded. The proposal by Mosonyi (1999) consists of the application of the Beta distribution function (whose main characteristics are presented in Table 6.1) – that is limited (bounded) on two sides. The initial value of the function is $u = Q_0$ when the discharge is higher then the carrying capacity of the riverbed, and the maximum value is $v = \text{PMF} = Q_{max,max}$.

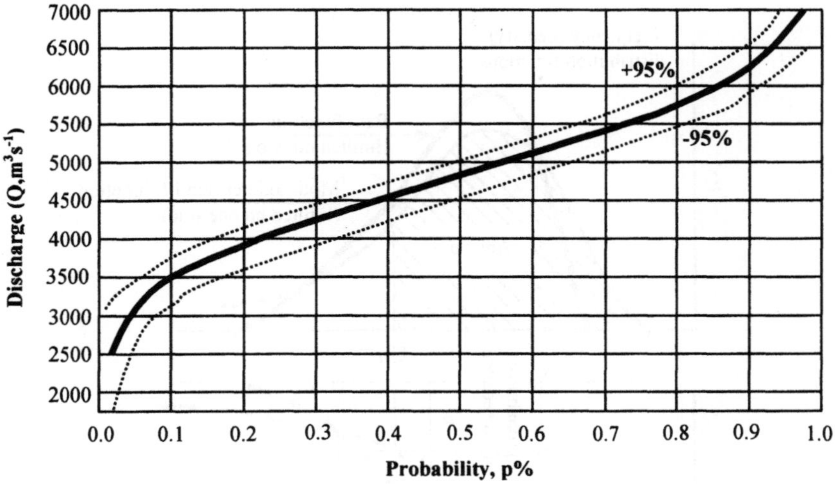

Figure 6.4. A schematic of a cumulative normal probability and empirical frequency distribution functions for design flood estimation for large rivers

Here, it is assumed that the values of maximum annual flood discharges, Q_1, Q_2,……,Q_n are known for an *n* time period, and that the empirical probability from the ordered sample of Q_i is $P_i = (i/n)$. In general, integration of the probability density function, $f(x)$, yields the cumulative distribution function, F(x) – with values of the latter lying between 0 and 1. Here the probability of non-exceedance is $F(x) = p$, and the probability of exceedance $F^*(x) = p' = 1 - p$.

Table 6.1. Basic characteristics of the Beta distribution function

The distribution function	$f(x) = C_o \cdot \frac{1}{(v-u)^{a+b-1}} \cdot (x-u)^{a-1} \cdot (v-x)^{b-1}$
Probability distribution function of non exceedance, $F(x) = \int_{z=0}^{x} f(x)dx$	$F(x) = B\left[\left(\frac{x-u}{v-u}\right), a, b\right]$ $B(x,a,b) = \int_{z=0}^{x} C_o \cdot z^{a-1} \cdot (1-z)^{b-1} dz$ $C_o = \frac{\Gamma(a+b)}{\Gamma(a)\cdot\Gamma(b)}$
Validity boundaries	$u < x < v$
Method of moments for estimating the parameters	$a = \frac{\bar{x}-u}{v-\bar{x}} \cdot b$, $b = \frac{v-\bar{x}}{v-u} \cdot \left(\frac{(\bar{x}-u)\cdot(v-\bar{x})}{s^2} - 1\right)$

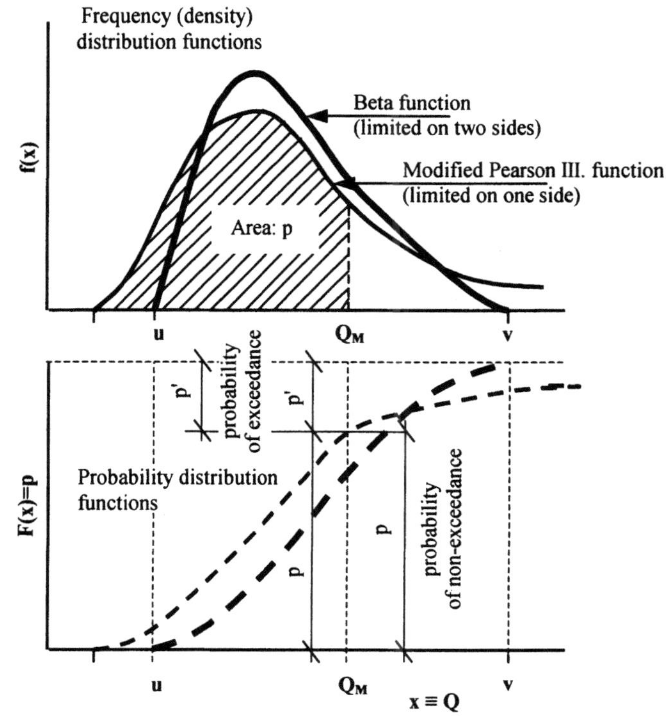

Figure 6.5. Types of distribution functions (after Mosonyi, 1999)

6.2.2. ESTIMATION OF THE PROBABILITY DISTRIBUTION FUNCTION OF THE MAXIMUM FLOOD FLOWS FOR SMALL RIVERS

The estimation of the peak flood discharge for small catchments areas is a rather difficult exercise. This is especially true because an extremely high intensity precipitation of usually very short duration can produce a high peak flood discharge – but a relatively small volume of water; on the other hand, precipitation with lower intensity can produce a relatively low peak flood – but a higher volume of water. As a consequence, the probability distribution of the annual maximum discharges of such rivers will not follow a normal distribution.

In general, the maximum values of local floods from the small rivers, Q_i^{max}, are independent random variables, viz.:

$$P(Q_i^{max}/Q_j^{max}) = P\,(Q_i^{max}) \qquad (6.11)$$

The probability distribution of extreme values of annual discharges can be described by the following relationship:

$$F\,(x) = P\,(NQ \leq x) = e^{-e^{-y}} \qquad (6.12)$$

Depending on the empirical distribution of the random variables, the exponent, y, in equation (6.12) can be related to a Gumbel-type distribution, viz.:

$$y = a(x+b) \qquad (6.13)$$

or, it can also can be related to a Frechet-type distribution, as follows:

$$y = A[ln(x) + B] \qquad (6.14)$$

In fact, to date there is no known theoretically based type of probability distribution function that adequately addresses local flood flows. Thus, in most engineering practice, we typically use graphic tests to facilitate the selection of a satisfactory distribution function. Typically, the procedure involves plotting the values of the annual maximum discharges in a Gumbel or Frechet system of coordinates; the type of distribution that produces a frequency distribution in which the plotted values are scattered around a straight line is then considered acceptable. In general, if the Gumbel type of probability distribution function is applicable, then the values of the random variables x should be nearly scatter around a straight line. On the other hand, when the Gumbel type of distribution is not applicable, then the applicability of the Frechet-type of probability distribution should be evaluated in similar manner.

Based on past practical experience, it has been determined that, for continental European types of climatic conditions, the probability distribution functions of the maximum annual discharges of small rivers and brooks can be characterized by the Gumbel or Frechet type of distribution – if the catchment area A < 10, 000 km^2. As an example, the results of fitting a Frechet-type of distribution is illustrated in Figure 6.6 using data from 36 years of observation records for the Karasica brook in Hungary.

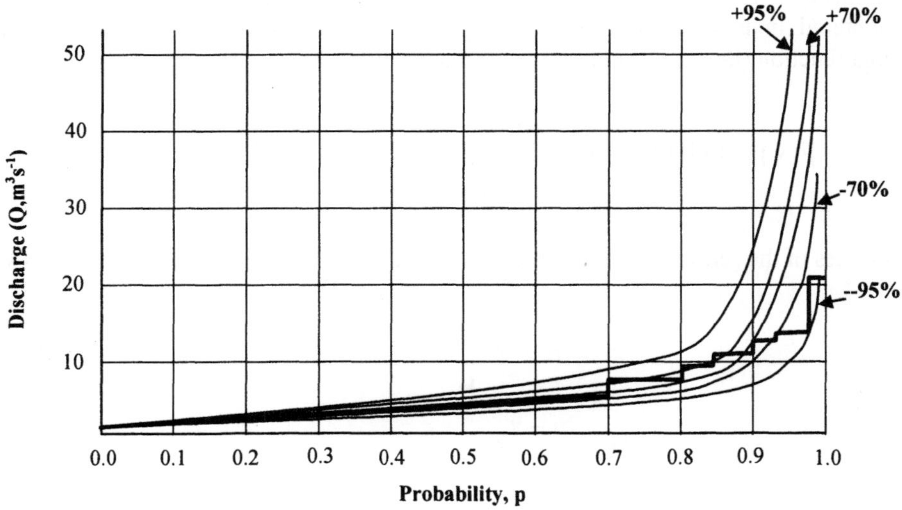

Figure 6.6. Illustrative example for the application of the Frechet type probability distribution function to design flood estimation for small rivers

6.2.3. ESTIMATION OF THE PROBABILITY DISTRIBUTION FUNCTION OF MAXIMUM FLOODS IN THE CASE OF MEDIUM-SIZED RIVERS

Medium-sized rivers are used to identify cases where the catchments area is between 10,000 km^2 and 100,000 km^2, i.e., 10,000 km^2 < A < 100,000 km^2. A common procedure for determining the maximum flood in the case of medium-sized rivers involves a graphical fitting of the measured data series to a Gaussian, Gumbel, or Frechet type of probability distribution function. Typically, the procedure involves plotting the values of the annual maximum discharges in a Gaussian, Gumbel, or Frechet system of coordinates; the type of distribution that produces a frequency distribution in which the plotted values are scattered around a straight line is then considered acceptable. The goodness-of-fit of the chosen theoretical distribution can be verified with the Kolmogorov test.

If the graphical analysis produces unsatisfactory results – as could happen if, for example, we have a long homogeneous time series – then one can investigate the maximum values of the discharges by using time intervals of 2, 3, or 4 years, and then repeating the graphical fitting of the data using the Gumbel or Frechet system of coordinates. It should be acknowledged that, if we use the maximum values from different time steps or units (e.g., 2, 3, or 4 years), then there is a consequential decrease in the accuracy of the evaluation. On the other hand, the standard deviation of the maximum values will decrease – and thus there is a decrease in the fluctuation of the data around the mean value. As an example to illustrate the effect of the use of a 2-year time interval, Figure 6.7 presents the results of a graphical test fitting for data from the Maros River in Hungary – using the Frechet probability distribution. The length of data series of annual maximum discharges that was used is 51 years.

If the length of measured time series is less than 30 years, then it is not possible to produce a reliable estimation of the characteristic parameters. In that case, it is recommended to use a traditional frequency distribution. Similar to what is encountered in much of this type of engineering problems, we very often use the gamma function (i.e., the Pearson Type III distribution), viz.:

$$F(x) = P(NQ \le x) \cong \frac{\beta^k}{\Gamma(k)} \int_{x_o}^{x} (t - x_o)^{k-1} \cdot e^{-\beta(t-x)} dt \tag{6.15}$$

The values of the relevant parameters can be calculated from the following expressions:

$$x_o = m - 2 \cdot \frac{\sigma^2}{\mu_3} \tag{6.16}$$

$$\beta = \frac{m - x_o}{\sigma} \tag{6.17}$$

$$k = \frac{(m - x_o)^2}{\sigma} \tag{6.18}$$

where m is the arithmetic average, σ is the standard deviation, and μ_3 is the third central moment of the statistical sample. Using the sample data to determine the x_o, k, β parameters will usually produce a good fit – that can be validated with the χ^2 test.

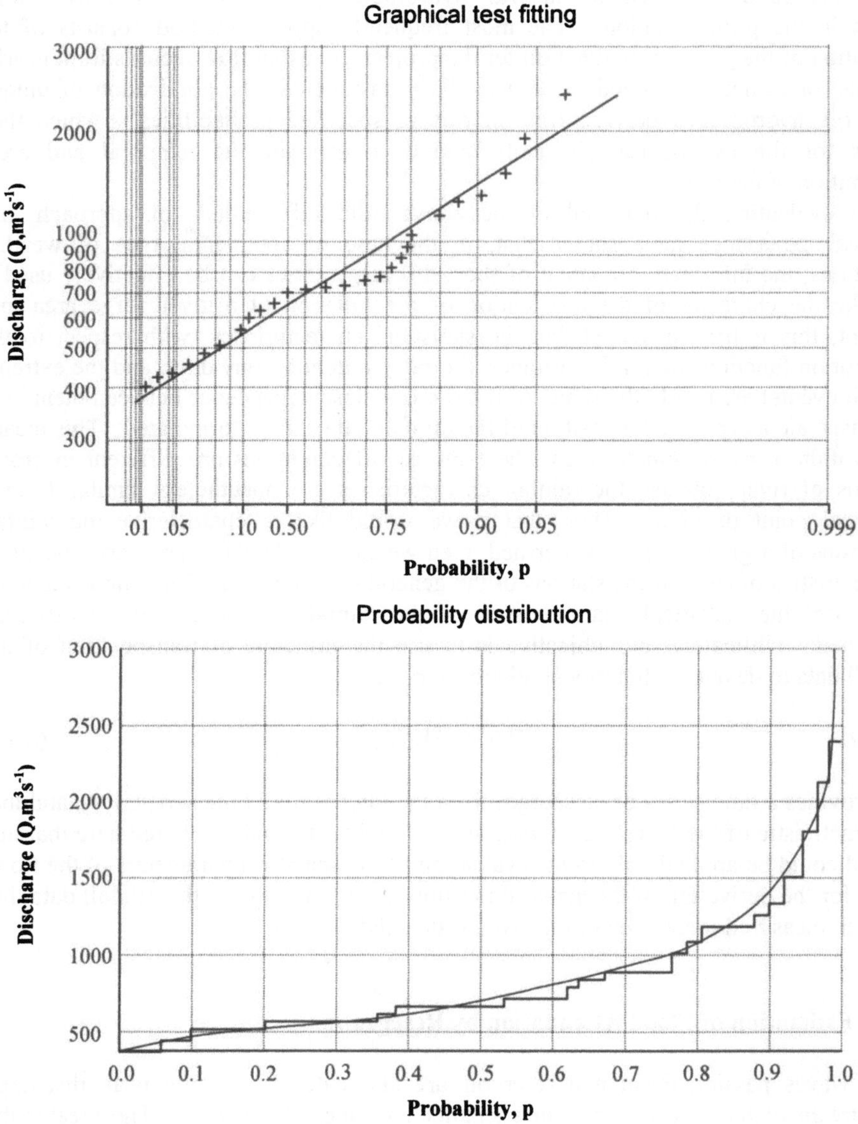

Figure 6.7. The Frechet type of probability distribution function for design flood estimation for medium-sized rivers

6.3. Estimation of Flood Flows Using Limited Data

It is apparent that, in the case of time series of $n < 10$ years, it is almost impossible to obtain reliable deterministic results about the probable distribution of floods – especially because these data usually will represent only a limited section of dry or wet weather periods. Under such circumstances, it becomes necessary to utilize synthetic statistical methods of approach – i.e., when the statistics of the probability function for the discharge data are consistent with hydrologically and meteorologically similar basins in the project region. The most frequently applied method consists of the derivation of the probability function for the project site using flow records from nearby streams, or at nearby gauged locations. This may include the application of unique discharge hydrographs derived from storms of specified probabilities – which then allows for the use of rainfall-runoff models to estimate the temporal and areal distribution of the flows.

In evaluating the potential of the above-indicated methods of approach, our practical experience have shown that, in the case of continental type of weather conditions, the information content of short discharge series can be effectively used if we take the character of the distribution of precipitation of a given large area into account; this is true, provided that the statistical characteristics (with respect to the distribution functions of yearly averages, the rainy and non-rainy days, and the extreme rainfall events) are nearly the same, or if there is a close dependence between them. On this basis, a concept of transposition of the rainfall data can be formulated. This means that, within a given climatic area where the runoff conditions are different in cross-sections of river courses, the rainfall characteristics are nonetheless similar from a statistical point of view. Thus, that if we would like to characterize the rainfall conditions of a given river or watershed, then we can use the data series and statistical characteristics of each of the stations of the general climatic area. This implies that we 'transpose' the statistical characteristics of the rainfall to the area of investigated watershed. Ultimately, our objective is to use the principle of transposition of the rainfall data to develop a functional relationship, i.e.,

$$Q_f(t) = A[P(t)] \tag{6.19}$$

that provides a time series of discharges from the long rainfall time series, P(t), and that is characteristic of the investigated watershed. Finally, it should be noted here that this method could be applied only to the evaluation of the statistical parameters of the flood flows; for the derivation of the actual flood flows, we can use only the rainfall data that has been measured in the area of the given watershed.

6.4. Estimation of Flood Attenuation by Reservoirs

Flood waves passing through a reservoir are attenuated – i.e., the peak discharge downstream of the reservoir becomes smaller then the inflow peak. The greater the capacity of the reservoir, the more pronounced this attenuating effect.

Reservoirs are basic subsystems in a water resources systems that can transform a flood wave, I(t), as the input into the outflow hydrographs, Q(t), as the output. The peak input discharge, $I_{max}(t)$, serves as a design factor in controlling water damage resulting from the peak output discharge, $Q_{max}(t)$, that is released to the tailwater section. The flood reduction rate of the reservoir, n, is defined by the following relationship:

$$n = \frac{I_{\max}}{Q_{\max}} \tag{6.20}$$

and this depends on the empty storage space available at the time of a flood, i.e.,

$$K_e = K - S(t) \tag{6.21}$$

The relationship between the ratio, n, and the empty space, K_e, depends on the reservoir operational schedule. It has been shown that, in order to reduce the inflow peak discharge, the empty storage space should be (Bukovszky, 1989):

$$S\left[Q_{\max} = \frac{I_{\max}}{n}\right] = A \cdot \frac{(n-1)}{n} \cdot V_{fl} \tag{6.22}$$

where V_{fl} is the entering flood volume beyond and above the base flow from groundwater into the river, Q_b, and the factor A is a constant that depends on the reservoir operational schedule. For continuously operated gates, A=1/2; for hydraulically well-designed spillways, A = 1; and for closed gates, A = 2.

Numerical calculations using equation (6.22) requires knowledge of the correlation of the peak discharge with the volume, V_{fl}, as follows:

$$I_{\max} = f(V_{fl}) \tag{6.23}$$

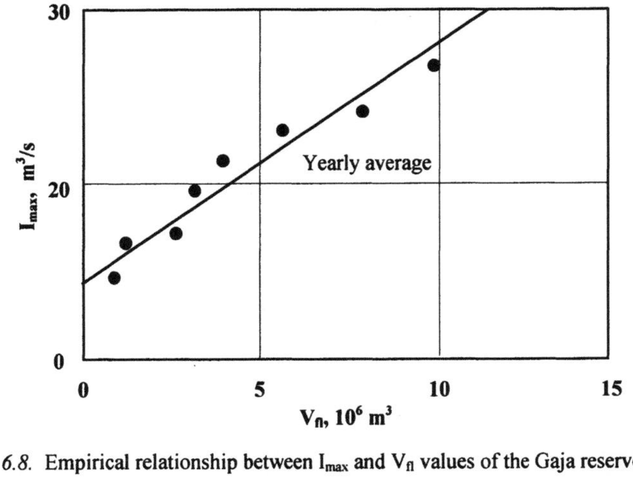

Figure 6.8. Empirical relationship between I_{max} and V_{fl} values of the Gaja reservoir in Hungary

This empirical relationship can be approximately evaluated using linear correlation analysis, as follows:

$$V_{fl} = \frac{I_{\max}}{b} \tag{6.24}$$

where the factor *b* may vary from season to season. This can be evaluated for the average conditions, as it illustrated in Figure 6.8 – showing the relationship between I_{max} and V_{fl} values.

Chapter 7

METHODS OF APPROACH FOR DESIGNING OPTIMAL STORAGE CAPACITIES AND OPERATIONAL STRATEGIES FOR MULTIRESERVOIR SYSTEMS

In the analysis of a reservoir development scheme, a number of alternative dam sites would typically be investigated, not only for the construction requirements but also from the hydrologic point of view. Procedures used for rapid assessment – usually designated as *preliminary design techniques* – involve making simplifying assumptions for ease of application; as a consequence, such procedures often result in a reduced accuracy. After using preliminary design techniques to eliminate unsuitable reservoir sites from further consideration, the remaining locations are evaluated by using a *final design technique*; these latter techniques tend to be more complicated and/or sophisticated because they take into account most, if not all, of the factors which influence storage (McMahon and Mein, 1986, 1978).

In general, the storage capacity necessary on a river to meet specified demands depends on numerous factors – such as the variability of the streamflows, the size of the demand, and the degree of reliability to meet the demand requirements. This chapter discusses some reservoir design techniques that may be utilized during either of the preliminary or final design stages in the hydrological dimensioning of storage reservoirs.

7.1. A Review and Classification of Reservoir Capacity-Yield Estimation Procedures

Reservoir capacity-yield procedures may be classified into three main groups, namely (e.g., McMahon and Mein, 1986, 1978):

(i) Critical Period Techniques;
(ii) Probability Matrix Methods; and
(iii) Procedures based on Stochastically-Generated Data.

Figure 7.1 shows a general picture of the reservoir capacity-yield classification methodologies in a very broad sense – with the specific key procedures elaborated further.

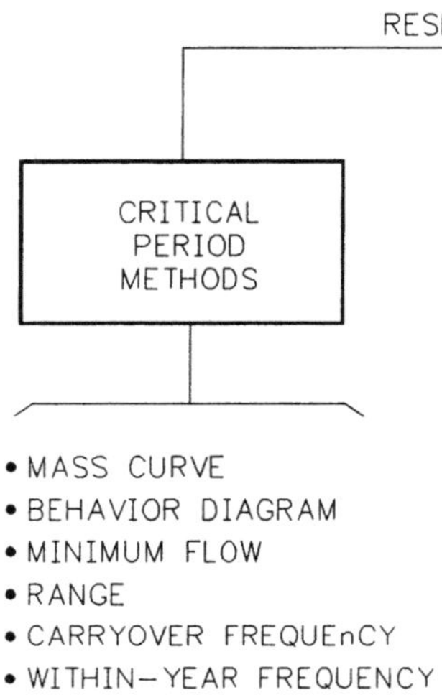

Figure 7.1. A classification of reservoir capacity yield procedures (adapted from McMahon and Mein, 1986; 1978)

7.2. Critical Period Techniques

Critical period techniques include methods in which sequence(s) of flow for which demand exceeds inflows is used to determine the storage size. There is no unique definition to the critical period. The U.S. Army Corps of Engineers (1975) define the *critical period* as the reservoir state from the full condition through emptiness to the full condition again; the state from fullness to emptiness is then referred to as the *critical drawdown period.* Under these circumstances and applying this definition, more than one failure would typically be encountered during the critical period. McMahon and Mein (1986, 1978) define a *critical period* as that during which a reservoir goes from a full condition to an empty condition without spilling in the intervening period; thus, under such circumstances, only one failure can occur during a critical period. A few other variations to these definitions to the critical period concept may indeed be encountered in the literature of reservoir design studies; however, it is less likely to have great deviations from the two presented above.

McMahon and Mein (1986, 1978) provide a review of the critical period methods of storage-yield estimation under the following three main headings:

(1) Methods which indicate reservoir fullness with time (i.e., the state of the reservoir with time);
(2) Methods based on the range of flows; and
(3) Methods based on low flow sequences.

Another possible classification is one based on whether the procedure allows storage to be related to probability of failure; indeed, those methods that ignore probability of failure are considered inadequate. In any case, close examination of the assumptions and theoretical basis of the critical period procedures reveals that they are all deficient in one way or another (McMahon and Mein, 1986, 1978).

The techniques based on the *state of the reservoir with respect to time* include the Mass Curve (Rippl Diagram) method and modifications thereof – such as the Residual Mass Curve Approach, and also the Behavior (or Simulation) Analysis and variations thereof – such as the Semi-infinite Behavior Analysis (McMahon and Mein, 1986, 1978).

The *range* is the basis of a number of important reservoir capacity-yield procedures, among which are Hurst's Procedure (Hurst, 1951), the Sequent-Peak Algorithm (Fiering, 1967; Thomas and Burden, 1963), and indeed a number of other related methods.

In a *minimum flow approach*, drought curves are constructed from the lowest sub-sequences of flows of varying durations from the historical record. A draft line is then superimposed, and the critical storage is estimated by a mass curve procedure. Alexander (1962) has developed a series of drought curves of different probabilities of occurrence, and from these derived generalized storage-regulation-probability curves.

7.2.1. RESERVOIR CAPACITY-YIELD ESTIMATION BY MASS-CURVE PROCEDURE

One of the earliest methods used for the estimation of the reservoir storage size to meet a given draft was *Rippl's method.* This technique is simple and widely understood, especially because the analysis uses historical data – but it also takes seasonality, autocorrelation, and other flow parameters into account. The assumptions are made that the reservoir is full at the beginning of the critical drawdown period, and that the future sequences of inflow will not contain a more severe drought than the measured discharge sequence. In fact, the assumption that the reservoir is full at the beginning of calculation can give misleading results when the record begins with a drought period. On the other hand, the method is readily adaptable to computer implementation – and can also allow for varying patterns of demand due to seasonal changes; a computer procedure, known as the *sequent-peak algorithm* (that is equivalent to the Rippl analysis) has been presented by McMahon (1993).

As an illustration of the mass-curve procedure, the general steps involved in the graphical method of estimation – for the case of an assumed constant supply – follows below.

1. For the proposed reservoir location, construct a cumulative curve of streamflow, preferably for monthly data. The result is a monotonous line having a horizontal section only in the periods when there is no inflow. The terminal point of the mass curve, (T,V), corresponds to the total volume, V, discharged through the investigated section during the whole period, T, of the observation (Figure 7.2).

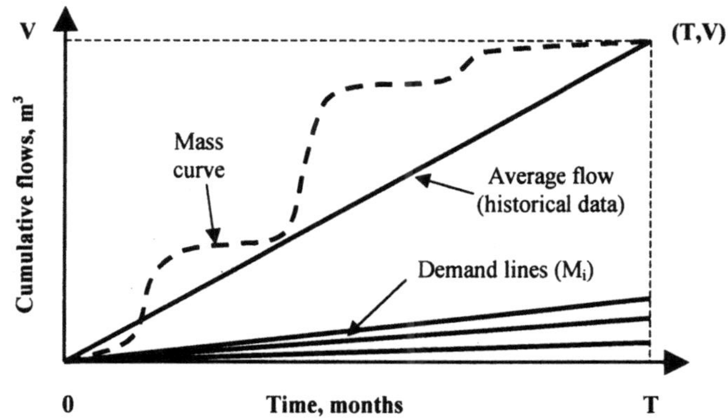

Figure 7.2. Mass curve and water demand lines

2. On the mass curve, superimpose the selected cumulative demand (supply) line (previously presented in Figure 7.2), such that it is tangential to the mass inflow curve (Figure 7.3). From Figure 7.3, it is apparent that the intercept K_2 is greater than K_1, and therefore the design capacity would be taken as K_2. The critical drawdown (water shortage) time period equals the maximum differences between the inflow and the water demand. It is noteworthy that, the demand lines can have a nonlinear (e.g., step-like) character, but the basic procedure involved remains the same.

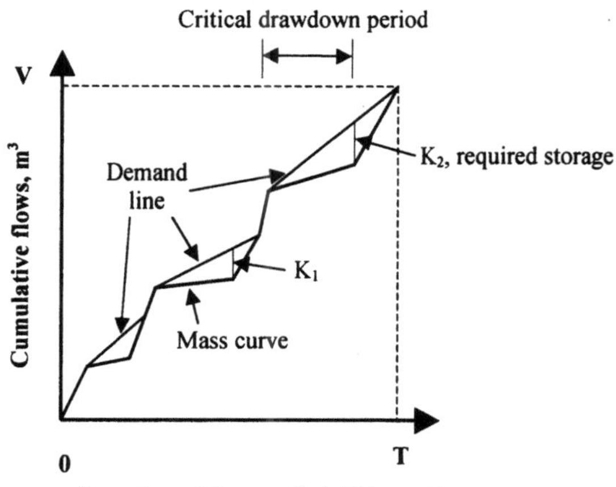

Figure 7.3. Reservoir capacity-yield estimation

3. It is recommended that the K_i values belonging to the different $M_i = Q_i(D_t)$ demands be determined first. Subsequently, construct a curve that represents the efficiency function, $K = f(M)$, for the given case – as depicted in Figure 7.4.

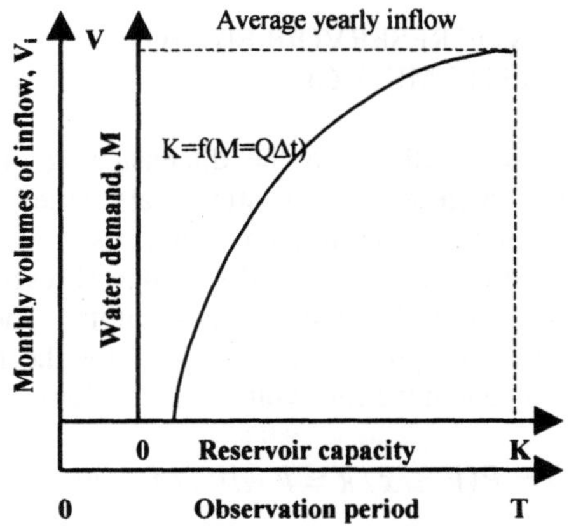

Figure 7.4. Efficiency function of the mass-curve methods

Overall, the mass-curve methods have several deficiencies when variable drafts, evaporation and other losses, and multi-reservoir systems are involved. Because of these limitations, the application of the mass curve method is best recommended only as a preliminary step in the reservoir design process – where only limited data exist.

7.3. Probability Matrix Methods

The probability matrix techniques, which constitute a very important group within the classes of reservoir-capacity estimation procedures, encompass the *Moran-related techniques* – mostly attributed to Moran (1959) – and their extensions and modifications thereof, such as the *Gould procedure*. From a theoretical point of view, the Gould procedure (Gould, 1964, 1961) apparently stands out as the most versatile reservoir capacity-yield technique under the probability matrix methods.

Essentially, the Gould procedure involves only one major assumption – and overcomes most of the disadvantages of other probability matrix procedures and the critical period methods. For these reasons, and because of the satisfactory results of several extensive programs (e.g., McMahon and Mein, 1978), the modified Gould procedure – that has been corrected for annual serial correlations – is recommended for use as a final design tool, in establishing the single reservoir capacity-yield-reliability relationship.

Indeed, a number of probability matrix methods exist in the literature, including the use of empirical equations for storage capacity estimation. One prime advantage of the probability matrix methods – and in contrast to the critical period techniques – is that, the computed storage estimates are independent of the starting conditions. Furthermore, non-continuous records can be handled more effectively. Also, since annual flows are assumed to be independent, the sequencing of flows becomes unimportant – and therefore records with missing data can be used as effectively as continuous records. Indeed, the procedures sample all years of streamflow records without reference to the historical sequencing – but with serial correlation assumed to be zero.

7.3.1. BASIC PRINCIPLES OF RESERVOIR SIZING USING PROBABILISTIC METHODS

The application of probabilistic methods to the dimensioning of storage reservoirs generally requires the use of a generated statistical data series and the theoretical probability distribution function of the random variables. It is on record that Savarenskij (1951) was the first to recognize that the change of reservoir volume should be analyzed using methods based on the theory of probability. The basic idea was to determine the series of conditional probability functions for the probability of non-exceedance of stored water volume in the reservoir, V – as follows:

$$F(x/k,m) = P(V \leq x/k = k \cdot dV, M = m \cdot dV) \tag{7.1}$$

where the condition is to satisfy the M = m·dV demands by using the K = k·dV reservoir volumes. Here, dV is the unit value of the volume, and *x*, *m*, *k* are parameter constants. By using this method, we can obtain exact result in the analysis of the joint processes of the independent random variables.

The dimensioning of reservoirs (characterized by capacity, K = k·dV and water demand, M = m·dV) involves the determination of the probability distribution of the water volumes in the reservoir. The random variable of interest – namely, the stored water volume in the reservoir – will have a restricted character because the lower limit is zero (equivalent to the emptying of the reservoir); the probability of occurrence of this situation represents the probability of breakdown of the water supply system, which is the measure of the calculated risk associated with the dimensioning of the reservoir. The upper limit of the random variable is the full volume of the reservoir (i.e., K = *k*), and the corresponding reliability value defines the probability of fullness of the reservoir (Figure 7.5).

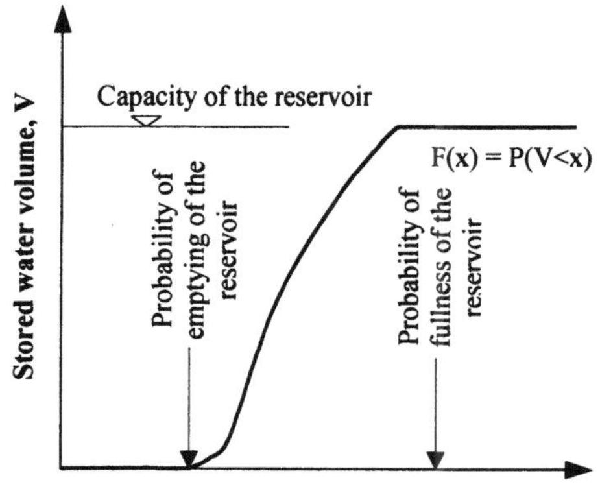

Figure 7.5. Typical probability distribution of the stored volume of water, V

The probability, P_o, of satisfying the demand M from the reservoir of volume K is equal to the conditional probability of emptying of the reservoir, viz.:

$$\begin{aligned} P_o &= F(x = 0 / K = k \cdot dV, M = m \cdot dV) \\ &= P(V = 0 / K = k \cdot dV, M = m \cdot dV) \end{aligned} \tag{7.2}$$

Thus, the demand M (= m·dV) from a reservoir with capacity K (= k·dV) can be satisfied with reliability, P, given by the following relationship:

$$P = 1 - P_o \tag{7.3}$$

Ultimately, this reliability of satisfying a given demand, M, from a reservoir with capacity, K, can be calculated by using a homogeneous system of linear equations containing 30-50 elements.

7.4. Reservoir Storage Requirements from Stochastic Data

The required capacity of a reservoir depends on the sequence of flows – especially the sequence of low flows. If a reservoir operates on an annual cycle (i.e., if it fills and partially or wholly empties each year), then it may be possible to assess its reliability on the basis of an analysis of the historical flow record – i.e., if this record is sufficiently long. If the reservoir operates with carryover storage (i.e., if it provides storage to meet required withdrawals over a period of several dry years), then the historical record is unlikely to provide adequate information on reliability. In most situations, records are generally too short to allow the probability of a series of 'sub-normal' years to be defined; stochastic methods provide an alternate means to estimating the probability of sequences of dry years during any specified period of a future time. Indeed, even if the historical records suggest that a reservoir will operate on an annual cycle, the possibility of two or more dry years following one another cannot be ruled out – and thus, stochastic analysis should preferably become part of the hydrologic study for all reservoirs that utilize natural stream inflows in the relevant analyses.

Storage estimation based on the use of generated or synthetic data involves using a stochastic generation model to 'manufacture' streamflow sequences with the same statistical properties as the historical record. Using some standard method in the follow-up analysis, it is then possible to determine the storage capacity corresponding to each sequence – thus providing the designer with a distribution of the relevant values. If a large number of sequences of streamflow are routed through a reservoir algorithm and the maximum storage requirement for each trace is determined, data for a probability analysis of storage requirements are made available. This in turn gives the designer an idea of the confidence that can be placed on the adopted design value. In any case, synthetic flows (i.e., stochastic flow data) do not improve poor records – but merely improve the quality of designs made with whatever record that is available. Broadly speaking, a stochastically generated data series will typically contain only the amount of information originating from the parent time series that formed the basis for the data generation algorithm.

7.4.1. BASIC ELEMENTS OF STOCAHSTICALLY-GENERATED DATA

Hydrologic design procedures are intended to provide more accurate estimates of the storage-yield relationship. Data generation is used to facilitate a better understanding of the sensitivity of a storage-yield estimate to other data sequences that could have been observed besides the historical one. In such efforts, the goal is to generate long (usually $\geq$ 1,000 years) data series that have the same statistical parameters as the observed data. This goal may be interpreted as the statistical 'caulking' of the information of a short observed data series such that, by utilizing the generated long data series having the same structure, it becomes possible to analyze the partial time series in which the features of the random variables remain the same. In this manner, the observed records are used for the estimation of the magnitude of the variable under consideration.

It should be recognized that the reliability of the generated data series depends on the information content of the original series. Consequently, various types of additional information (e.g., records of precipitation, infiltration, evaporation, and soil moisture conditions) can be used to complement the measured discharge data series.

Extension of the Monthly Discharge Data Series Using Precipitation Records

As an example in the case of the reservoir dimensioning exercise, the time series of monthly discharges may be estimated by using the monthly precipitation data (Loucks *et al.*, 1981). The basic approach to data series extension, as successfully applied in practice, is the use of the cross-correlation relationship – which can be derived by a linear regression between the precipitation and discharge data sets. This implies using the multivariate 'partial' cross-correlation method based on the principle of least squares.

Depending on the extent and the geological structure of the watershed, the influence of the precipitation on the discharges of the watercourse may be one of a delayed response – and this effect can be unequivocally demonstrated by using cross-correlation analysis. The synchronizing level between the monthly precipitation data and flow volumes is determined in accordance with the rules of the method of linear regression. The first step is the determination of the relationship between the corresponding data – as follows:

$$V(i) = a_o C(i) + b_o \tag{7.4}$$

This relationship is characterized by the 'single-step' coefficient of correlation for the two variables, $r_o(C,V)$. The next step in the analysis is the determination of the relationship between the sum of monthly flow volumes and the sum of precipitation of the previous month, in accordance with the following linear relationship:

$$V(i) = a_1 C(i-1) + b_1 \tag{7.5}$$

This can be used to determine the relationship between the shifted data and the corresponding coefficient of correlation, $r_1(C,V)$. Continuing with the above-described procedure would produce values for $r_2(C,V)$, $r_3(C,V)$.........$r_n(C,V)$. A graphical plot of such information gives the cross-correlation function, $r_n(C,V)$, as a function of the lag time, *n*; this is illustrated in Figure 7.6 for a small Hungarian stream, Kigyos, using 12 years of available data.

As a first approximation, the maximum point of the cross-correlation curve will determine the length of the lag time from where the previous precipitation data can be used for the estimation of the monthly flow volumes. Naturally, the character of the function and the corresponding lag time may be different along the river stretch – and indeed depends on the seasonality of the precipitation, runoff, and infiltration conditions. Hence, it is recommended that a series of cross-correlation functions be determined – and then investigate the effect of the previous monthly precipitations (that have different lag times) on the monthly flow volumes. This means that, it is best to analyze the above relationship (to include an evaluation of the effects of precipitations from the previous months on the given monthly flow volumes) by using a linear partial cross-correlation function (having *n* variables), as follows:

$$V(t) = a_n + b_{n,o}C(t) + b_{n,1}C(t-1) + b_{n,2}C(t-2) + \ldots + b_{n,i}C(t-i) + \ldots + b_{n,n}C(t-n) \tag{7.6}$$

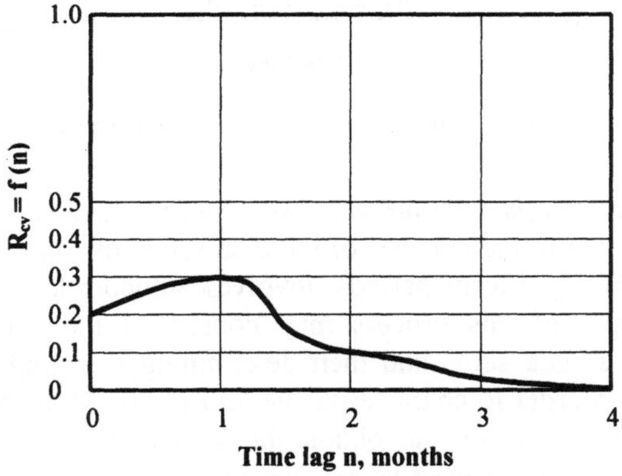

Figure 7.6. Schematic of the cross-correlation function

The first step is the evaluation of the 'partial cross-correlation' relationship is the determination of the coefficient of correlation, $R_o[V(t),C(t)]$, of the simple bi-variant relationship. Extending the estimation by taking the precipitation values of the previous months into account produces the multiple coefficient of correlation, $R_i[V(t), C(t), C(t-1)\ldots C(t-i)]$. This represents a subjective choice of the coefficient, $R_n[V(t), C(t), C(t-1)\ldots C(t-n)]$, that shows the maximum span of time after which the effect of the previous precipitation is negligible. In general, the fractional cross-correlation curve, $b_{n,n} = f(i,n)$, can be used to determine the length of the memory of the system – i.e., the maximum length of time during which the previous precipitation can influence the values of the flow (Figure 7.7). If $b_{n,n} < 0.1$, then the last value is negligible.

As has been established in the literature, it is noteworthy that by using the Fourier method of analysis, a continuous function can be presented as a series of sinusoidal waves. Assuming here that the continuous curves of the time series of discharges can also be characterized by univocal series of sinusoidal waves, it can be demonstrated that only the significant sinusoidal waves will represent the physical basis of the time series of discharges – and consequently they can be used to characterize the observed and stochastically generated information.

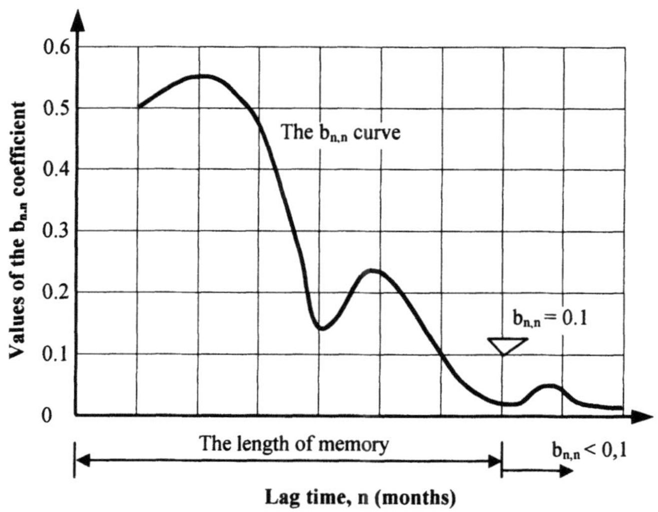

Figure 7.7. Interpretation of the fractional cross-correlation curve

Generation of Long Daughter Series with Identical Structure to Parent Time Series
An important first step in the analysis of the data series of water volumes, q(t), and the determination of the significant periods, involves autocorrelation analysis of the time series. An example of this process may consist of the determination the linear correlation among a data set – and then determining the single step autocorrelation coefficient, $r_{x,x}(1)$, in order to characterize the 'closeness' of the relationship. Then, by continuing the procedure, we can obtain the series of the steps of values, $r_{x,x}(2)$, $r_{x,x}(3)$.......$r_{x,x}(\tau)$ that can be used to develop the autocorrelation function of the flow volumes, $r_{x,x}(\tau) = f(\tau)$; this is illustrated in Figure 7.8 using the 17 years of data for the average monthly flows from the Mohacs section of the Danube River in Hungary.

The periodic effect of the annual precipitation on the generated discharges is quite obvious from Figure 7.8. In fact, when the autocorrelation function of the yearly or average monthly discharges, $r_{xx}(\tau) = f(\tau)$, exhibits a fluctuating character, then it is quite possible that periodical components are also present. For a more detailed analysis of the periods, one can calculate the Laplace transform of the autocorrelation function (spectrum); for the calculation of period length near the local maximum points (determined by the spectrum), the approximate Labrouste analysis can be used. The amplitude and phase belonging to the fixed periodic lengths of the sinusoidal waves can be determined by applying the numerical methods of the Fourier-type expansion, with the help of an appropriate computer algorithm.

On the basis of the above type of analysis, it is possible to determine one or more significant periods that will characterize the data series and its structure. Subsequently, the ordinates of the significant periods are subtracted from the observed (or extended) data series – resulting in the data series of the aperiodic residue. The result is the data series of the aperiodic residue, given by:

$$q^{\otimes}(t) = q(t) - \sum_{t=1}^{c} A_i \sin(b_i t + c_i) \tag{7.7}$$

where A_i, c_i and b_i are, respectively, the amplitudes, phases, and relational coefficients. A schematic representation of the overall procedure is provided in Figure 7.9 – depicting the time series of the precipitation; the generated part of the discharges using the precipitation data; the observed discharge data; and the significant period, P_1 (with an amplitude of A_1, and phase of c_1). The random aperiodic function, $q^x(t)$, having period P_2, amplitude A_2, and phase c_2 is derived next.

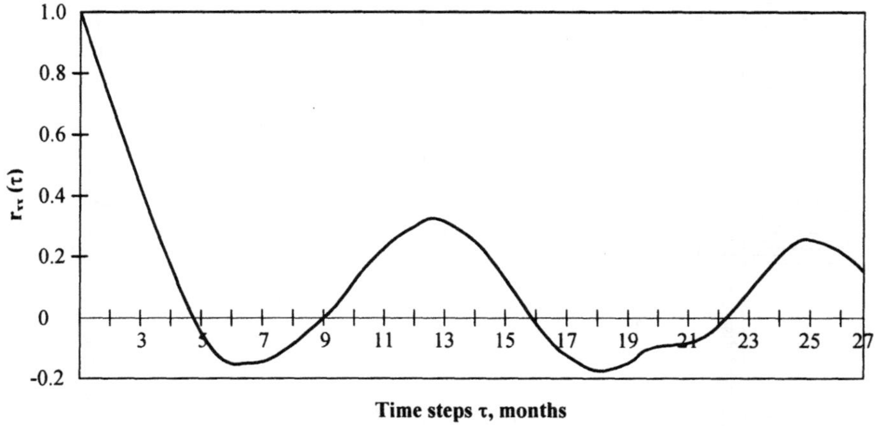

Figure 7.8. The autocorrelation function of the average monthly values of discharges

Further yet, it is possible to extend the data series (having a length of N years and the same probability distribution) by using Monte Carlo analysis. Application of this method requires the frequency distribution curve of the elements of the $q^x(t)$ process – given by (Figure 7.10):

$$G(x) \approx P\lfloor q^x \leq x \rfloor = F(x) \tag{7.8}$$

The last and relatively simple step of the data generation consists of the addition of the ordinates of the sinusoidal waves of the significant periods to the periods of the aperiodic elements. By using the long data series of the generated inflow and the water balance equation of the 'simulated' reservoir, and by analyzing the frequencies of different components of the operation, the series of curves that represent the characteristic capacity function can be derived, according to the following relationship:

$$K = f(M, R \approx P) \tag{7.9}$$

The practical application of the outlined procedure is quite simple if we use a suitable computer algorithm – albeit some new difficulties arise if we have to perform the analysis in relation to the joint behavior of several random variables. For example, if we have to take the evaporation losses of the reservoir or the random decrease of the flood waves into account, or if the task involves the analysis of a system of reservoirs, then the practical applicability of this method is called into question.

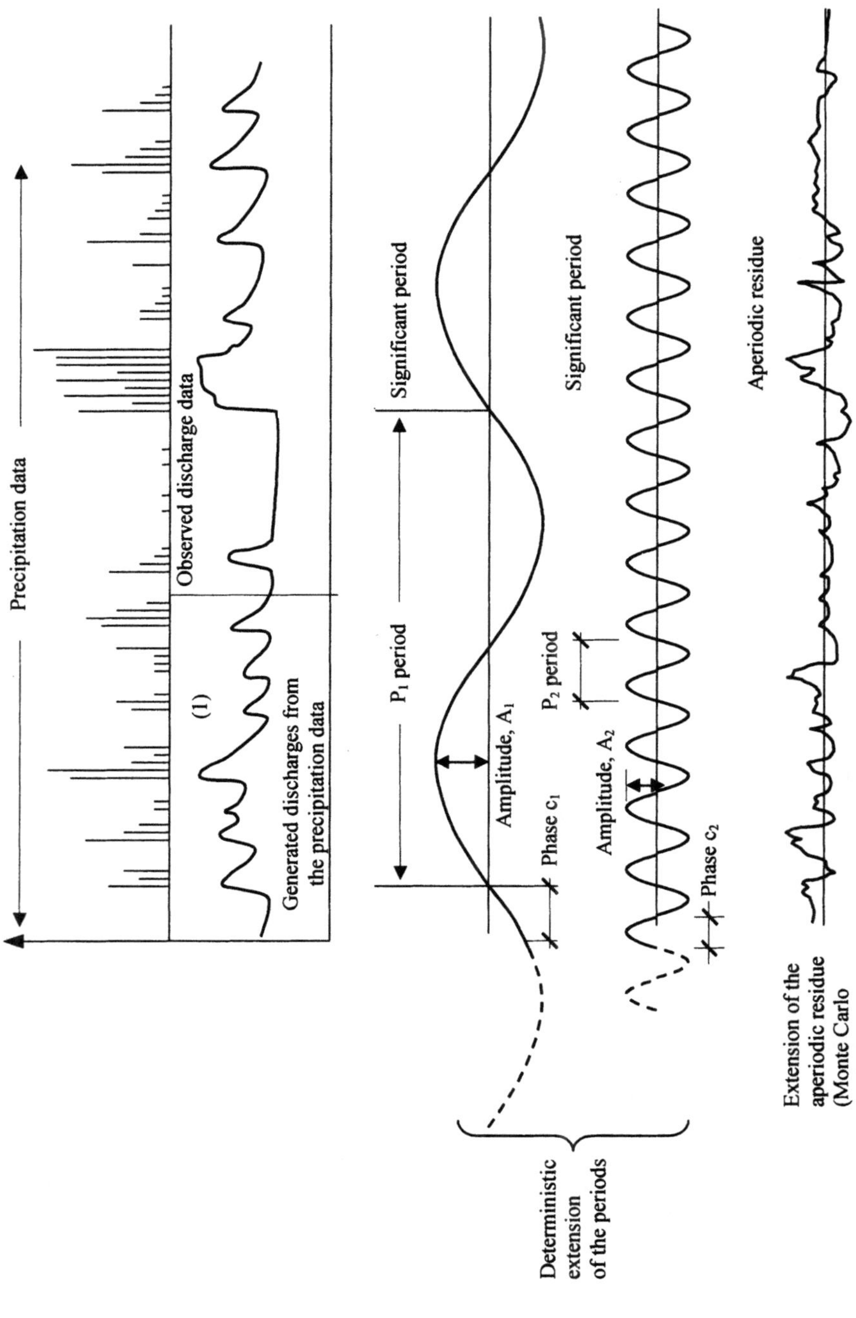

Figure 7.9. Illustration of the data generation scheme

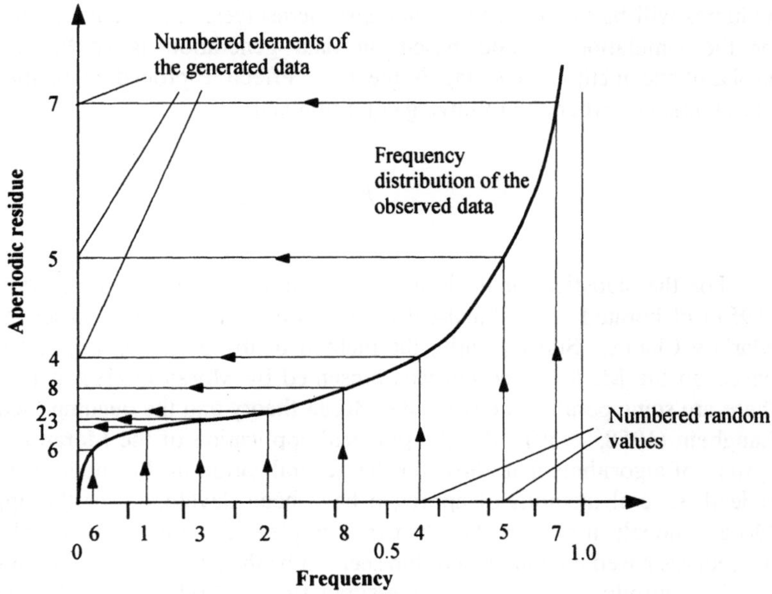

Figure 7.10. Using the Monte Carlo method in a data generation scheme

7.5. Choosing Between Deterministic vs. Probabilistic vs. Simulation Methods

The concept of *reservoir relative capacity* may be used to facilitate the choice of the best hydrological dimensioning technique to utilize in the reservoir capacity allocation process. The relative volume (capacity) of a reservoir is represented by the following relationship:

$$k = \frac{K}{V} \tag{7.10}$$

where K is the volume of the reservoir, and V is the multi-annual average (i.e., the expected value) of the yearly inflows. In the case of small reservoirs, i.e., where:

$$k = \frac{K}{V} \leq 0.2 \tag{7.11}$$

the dimensioning of the reservoir by use of probabilistic methods is *not* recommended. In such cases, the Gould method can be utilized; also appropriate is the application of the simulation method based on data generation as well as the Rippl method (using the mass curve for weekly discharges). In the case of reservoirs, where:

$$0.2 \leq k = \frac{K}{V} \leq 0.6 \tag{7.12}$$

(especially as found in karstic regions), it may happen that the half-yearly water volumes will be interdependent amongst themselves. Consequently, the Gould method or the simulation method based on data generation is applicable. Finally, the probabilistic method of sizing is the most effective procedure in the case of 'large' reservoirs, i.e., when the following condition is true:

$$k = \frac{K}{V} \geq 0.6 \tag{7.13}$$

For the statistical analysis of the stored water volume in reservoirs, Savarenskij (1951) elaborated a graphical-calculation procedure that was based on the theory of Markov Chains. Subsequently, the method of the dimensioning of storage reservoirs based on the Markov-theorem was presented by Moran (1959); it is noteworthy that, there are some parallels between the Moran theory and the queuing theory proposed by Langbein (1958, 1959). For the practical application of the Moran model, a complex system of algorithms that allow for the generalization of this model is usually required. Indeed, several methods of approach have been developed for the application of the Moran model; however, faced with complex computational problems, simulation procedures based on data generation seem to be the preferred method of choice.

The introduction and use of the simulation methods are based on the deterministic traditional mass-curve procedures. Unquestionably, this method can be applied in the case of limited and incomplete data – and for this reason, the possible application of this method may be considered quite extensive.

Finally, it is noteworthy that, an observed data series can be used for the statistical characterization of the one-dimensional distribution of the population only if the elements are *homogeneous* and also *independent* from one another. For the *testing of homogeneity*, the Kolmogorov-Smirnov or Wilcoxon two-sample test is usually employed. Other variant methods to utilize may include the Mann-Whitney test (Lehmann, 1975); the combinatorial method – which is based on the generalization of the Gnedenko-Kololyuk test (Reimann, 1989); and the stepping combined Smirnov-Kolmogorov test (Goda, 1995). In the *test of independence*, a special variation of the autocorrelation analysis can be used – alongside other classical methods. The essence of the method is the calculation of the single-step autocorrelation coefficients between water volumes belonging to the different time intervals – and these coefficients are presented as a function of the time steps, viz.:

$$r_{xx}^{\Delta t}(\tau = 1) = f(\Delta t) \tag{7.14}$$

The mean discharges and water volumes for a given time interval, Δt, are relatively independent from one another if,

$$r_{xx}^{\Delta t}(\tau = 1) = f(\Delta t) \leq 0.2 \tag{7.15}$$

The values are dependent when,

$$r_{xx}^{\Delta t}(\tau = 1) = f(\Delta t) > 0.3 \tag{7.16}$$

and the independence is uncertain if:

$$0.2 < r_{xx}^{\Delta t}(\tau = 1) = f(\Delta t) \leq 0.3 \tag{7.17}$$

Chapter 8

DETERMINATION OF THE OPTIMAL RESERVOIR STORAGE CAPACITY AND OPERATIONAL PARAMETERS FOR A RIVER DAM

Reservoir systems for flow regulation generally involve highly dynamic decisions. Of particular interest, the *optimal system design* may be defined as that combination of system units/components, inputs, outputs, and operating policies that 'extremizes' (i.e., maximizes or minimizes, as appropriate) the objective function – thus fulfilling the overall objective better than any other combination. The problem of finding the optimal capacities for the components in multireservoir systems, especially in situations where reservoirs are acting both in parallel and in series, becomes computationally complex due to the rather large number of feasible configurations that need to be analyzed. Typically, mathematical programming – involving mathematical optimization techniques – guarantees an optimal solution to this type of problem. However, because the problem formulation only approximates the real problem, the optimal solution obtained by mathematical optimization techniques usually is only an approximation of the optimal strategy for the system. Nonetheless, a good solution – including an analysis of the sensitivity of the solution to changing conditions – may be derived less expensively in this manner.

In general, a tradeoff between modeling perfection and computational tractability confronts all multiobjective water resource planners and analysts/designers. Thus, the modeler who does not compromise will not be able to find optimal policies, and the theorist too preoccupied with his/her computation will obtain an optimal policy to an unrealistic problem. A compromise solution should therefore be sought/attempted for most problems.

In view of the fact that the optimum design of any system is dependent not only on objectives and specifications, but also on the probable operating policy, the optimal design of reservoir systems normally must follow the selection of the probable target outputs and operating policies. A mathematical model that depends on its purpose and/or use is necessary to optimize the quality of design – and this must be specified very clearly.

8.1. Screening Models for Multireservoir Systems Design

Simple optimization models usually are important tools in the identification of potentially attractive reservoir system configurations and capacities during the preliminary stages of a system design. Optimization models used in the selection of an optimal range of alternatives – rather than the specific optimal alternative – for further analysis by more realistic models are called *screening models*. Following the use of such screening models to identify cost-effective and potentially efficient system configurations and component capacities, these designs can subsequently be refined and evaluated with the use of detailed *simulation models* (see, e.g., ReVelle, 1999; Mays, 1996; Wurbs, 1996; Jacoby and Loucks, 1972; Stedinger *et al.*, 1983). The primary function of screening models is to help eliminate undesirable project alternatives – prior to the application of a detailed simulation model to the desirable alternative(s).

Several families of multireservoir screening models have been developed – amongst them, the *explicitly stochastic reservoir model* (Thomas and Watermeyer, 1962; Loucks, 1969; Jacoby and Loucks, 1972; Houck and Cohon, 1978; Loucks *et al.*, 1981) and the explicitly stochastic models based on *linear decision rules* (LDRs) and *chance constraints* on release and storage volumes (ReVelle *et al.*, 1969; Loucks, 1969; ReVelle and Kirby, 1970; Nayak and Arora, 1971; Houck, 1979; Stedinger *et al.*, 1983). An alternative screening model design incorporates implicitly in a model, the probability distribution of natural unregulated flows by having the model's representation of system performance depend on either a historical or an average streamflow sequence; *linear-programming* (LP) models had been employed in such types of studies (e.g., Maass *et al.*, 1962; Thomas and ReVelle, 1966; Major and Lenton, 1979). *Dynamic-programming* (DP) models of system operation make inefficient screening models because they do not identify optimal system component capacities and targets directly; nonetheless, they have found application in some studies to the optimization of the reservoir sizing problem, and also to set hydropower targets (e.g., Hall and Howell, 1963; Trott and Yeh, 1973; Loucks *et al.*, 1981).

A basic problem with *implicitly stochastic models* is that, many periods may need to be included in a model if an adequate implicit definition of the distribution of unregulated natural streamflows is to be obtained. This can be avoided, in part by designing for the 'Critical Period' of record (e.g., Hall *et al.*, 1969; Trott and Yeh, 1973). *Stochastic LP design models* are appropriate for analyzing relatively small multireservoir river basin systems having streamflows that are highly cross-correlated.

The *yield model* developed by Loucks *et al.* (1981) is a general purpose, implicitly stochastic LP screening model that incorporates several approximations to reduce the size of the constraints needed to describe reservoir system operation, and to capture the desired reliability of target releases. Thus, this can be used to estimate the storage capacities required to deliver various releases (yields) with specified reliabilities. The model estimates separately, over-year and within-year reservoir capacity requirements to meet specified release-reliability targets. The yield model is indeed quite flexible, and can be used with different assumptions to generate prospective cost-efficient designs with release reliabilities in the neighborhood of a target.

A *reliability programming* approach to the design of multipurpose reservoir systems – that allows for the direct inclusion of risk in an optimization analysis – has been considered by Simonovic and Marino (1980, 1982). The reliability programming technique – incorporating the concept of risk or reliability in an optimization procedure

– is applied to a multipurpose system of reservoirs, in which each reservoir could be connected to every other reservoir, and also each could receive releases from any or all other reservoirs as made necessary by the particular system configuration; the reservoirs can be connected in parallel or in series, and with linkages being normal channel flow for reservoir releases. Each reservoir in each period could receive random inflows, regulated inflow from upstream reservoir releases, inflow from pumping, diversion, etc. The outputs are random and deterministic demands (such as irrigation requirements, municipal and industrial water supply); evaporation and seepage losses; and pumping and diversion to other reservoirs and/or regions. The system can then be described by the state of the reservoirs, in accordance with the principles of continuity, viz.:

$$S_i(t)=\sum Q_i(t)-\sum R_i(t)-\sum_j \varepsilon_i S_i(t) \tag{8.1}$$

$$S_{i,\min} < S_i(t) < S_{i,\max} \qquad i=1,2,........,n; t\in[0,T] \tag{8.2}$$

where $S_i(t)$ = storage in the *i*-th reservoir; $Q_i(t)$ = inflow into the *i*-th reservoir; $R_i(t)$ = release from the *i*-th reservoir; $\varepsilon_i S_i(t)$ = evaporation loss from the *i*-th reservoir (as a fraction of its storage); n = number of reservoirs in the system; T = end of the simulation horizon, with the time domain being from 0→T; and $S_{i,max}$, $S_{i,min}$ are the maximum and minimum 'permissible' storage, respectively, in reservoir *i*. The reliability programming formulation includes an objective function based on economic efficiency representing the tradeoff between benefits and risks – i.e., maximization of the difference between net benefits and risk losses. Henceforth, the necessary mathematical model can be formulated on this basis.

Stedinger *et al.* (1983) provides a comparison of the ability of several of the screening models to identify efficient combinations of reservoir design capacities – illuminating some of the relative advantages and limitations of the methods with an appropriate sample problem; the models considered included a LP model based on mean monthly flows, a LP model based on critical period flows, the yield model, and LDR chance constrained models that use two release rule structures. It is noteworthy that, variations of these procedures either exist or can be formulated for reservoir systems designs.

Overall, screening models include several assumptions – such as hydrologic determinism – that can render such models relatively unreliable, especially due to the overly optimistic viewing of the world that is built into the model vis-à-vis the fact that river basin systems are characterized by a high degree of hydrologic stochasticity, and that streamflows are inherently uncertain processes.

8.1.1. A COST-EFFICIENT RESERVOIR CAPACITY DESIGN IN MULTIRESERVOIR SYSTEMS

Invariably, the key problem in all reservoir systems design is to determine a cost-efficient reservoir capacity for each site in question – such that the resultant reservoir system can provide the target release/demands with the desired performance reliability. To develop a model that identifies cost-efficient reservoir capacities, K_i, it would be necessary to specify reservoir capacity cost functions, $C(K_i)$, for each potential reservoir site *i*; such cost functions may take varying forms, such as:

$$C(K_i)=[K_i]^{\alpha} \quad or \quad C(K_i)=f(K_i) \qquad \forall i \tag{8.3}$$

where α is a constant real number, and $f(K_i)$ denotes some function of K_i. Subsequently, the formulation of the overall system model objective would be stated as the maximization of the net benefits that can be derived from the project. Alternatively, the formulation may consist of the minimization of the total annual cost of the system, less the annual benefits from the system, being used as the objective function. From a pre-analysis point of view, the model's parameters and the resultant design would consist of random variables – and a stochastic streamflow model may be used to represent the actual/historical multivariate distribution of flows at the reservoirs sites.

For a simple reservoir configuration and demand pattern, the *space rule* (Maass *et al.*, 1962) provides an efficient and reasonable system operating policy that can be used to evaluate all designs. The space rule allocates empty space among the reservoirs so as to minimize the probability of spills; this is achieved, insofar as possible, by making the empty space in each reservoir proportional to the expected inflow into that reservoir.

8.2. Reservoir Network Analyses for Model Development

The specific objective in model development is, almost invariably, the extremization (i.e., maximization or minimization, as the case may be) of one or more objective functions, subject to a set of constraints. Given a set of exogenous input variables – many of which are likely to be uncontrollable (e.g., precipitation or streamflow) – and given the system relationships (e.g., production function described by storage-yield curves, and sociopolitical constraints imposed by water rights laws, both of which contain controllable and uncontrollable variables), we may want to select or determine the system components (comprising of reservoirs sizes, for instance) and to select design parameters for these components (e.g., reservoirs storage allocation among various uses) in order to maximize the value of the concomitant endogenous output variables (e.g., water supply, water quality, etc.) less the input resource costs. The identification of the system parameters – called systems modeling – is indeed essential to arriving at an optimal design.

In formulating the design model, certain general and specific assumptions are made for ease of manipulation. In any case, the basis for developing an optimization model for the design of multireservoir systems with arbitrary configuration would comprise a consideration of a hypothetical but representative system – as depicted in Figure 8.1. Figure 8.2 then provides a three-level network representation of the spatial and temporary linkages between the system components at different simulation time levels. The basic process in each reservoir is the storage and allocation of water at every stage of the simulation; the change in storage during the *t*-th period of the τ-th year of regulation is defined, in discretized time intervals, as follows:

$$\left.\begin{matrix} S(\tau,t+1) \\ S(\tau+1,t) \end{matrix}\right\} = \max\{\min[S(\tau,t)+Q(\tau,t)-R(\tau,t),K],0\}_{t=T}^{t=1,2......(T-1)} \tag{8.4}$$

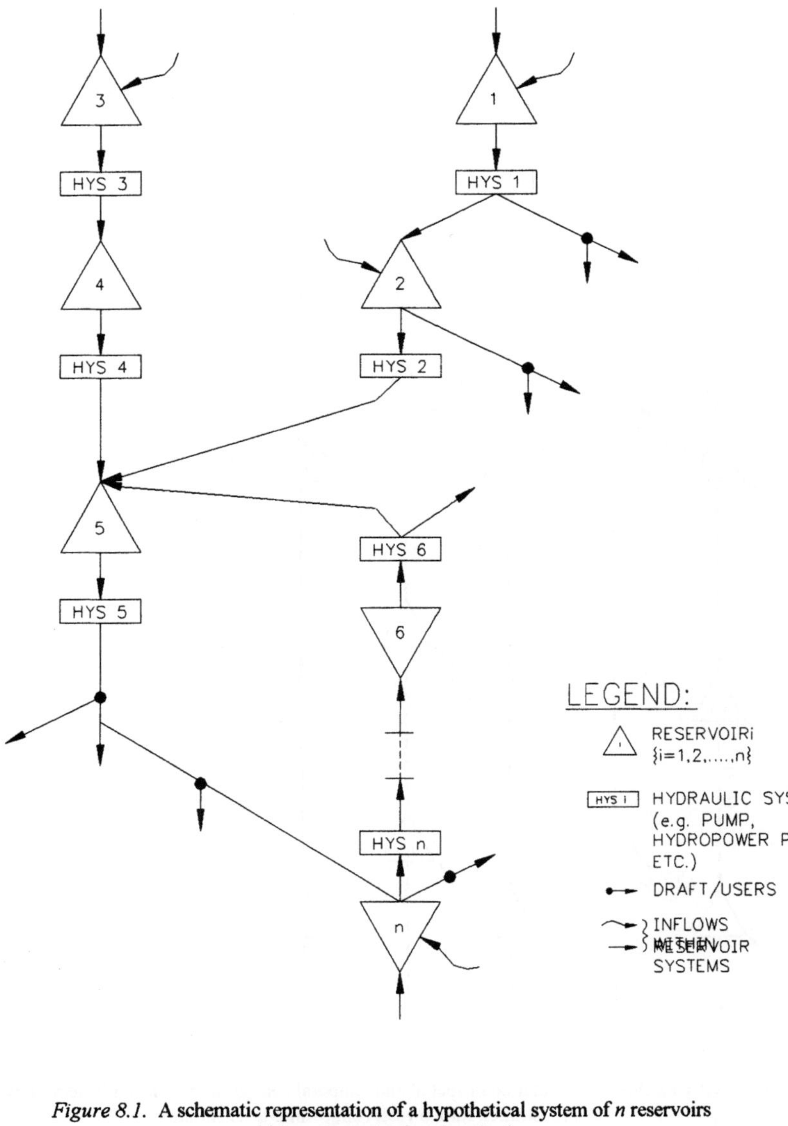

Figure 8.1. A schematic representation of a hypothetical system of *n* reservoirs within a representative river basin

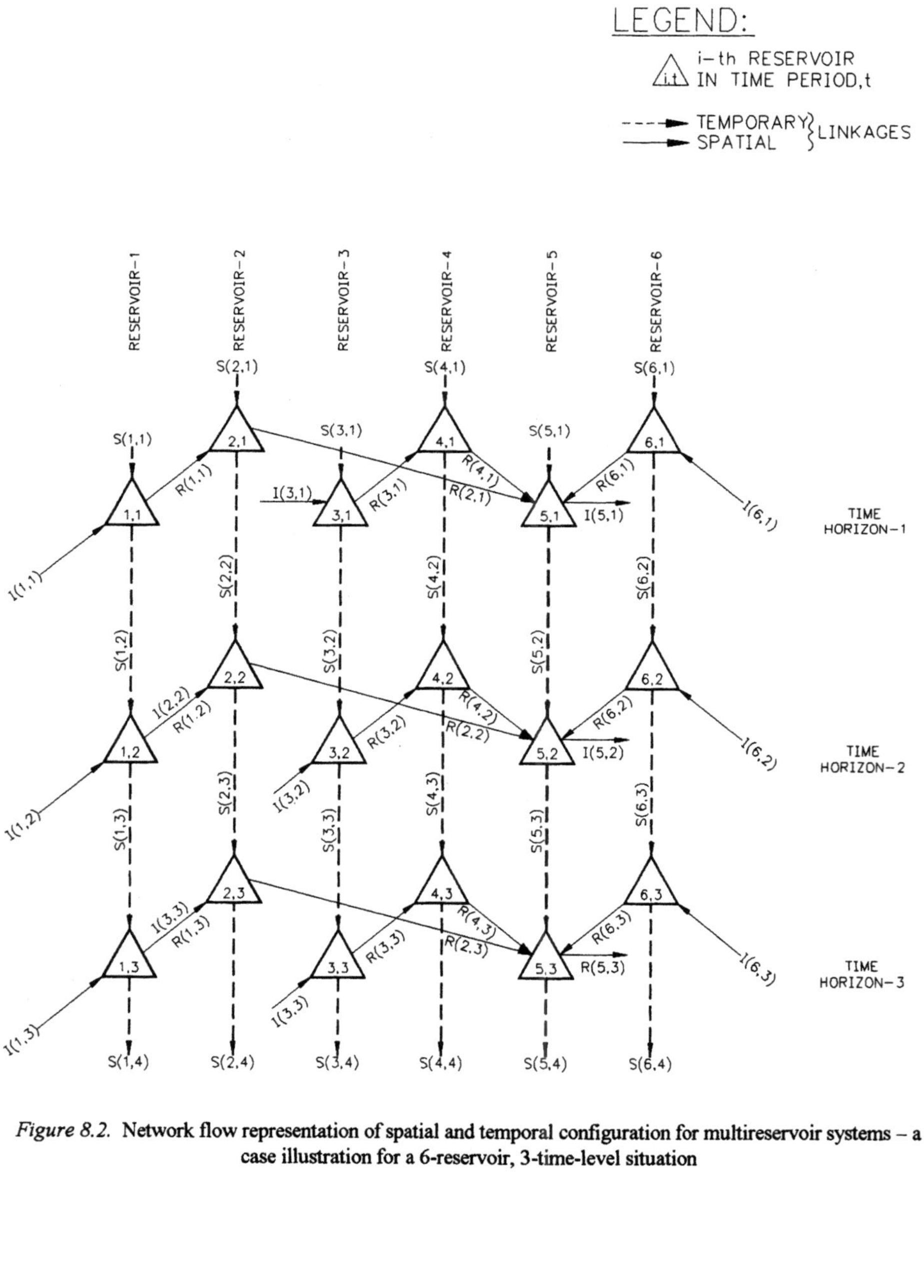

Figure 8.2. **Network flow representation of spatial and temporal configuration for multireservoir systems – a case illustration for a 6-reservoir, 3-time-level situation**

where $S(\tau,t)$ = storage at the beginning of period t in year T; $Q(\tau,t)$ = inflow during period t in year τ; $R(\tau,t)$ = release during period t in year T, defined by the release rule for any time interval which is a function of the operating policy; K = capacity available for regulation; and T = simulation time domain. In general, the following relationship holds true for the i-th reservoir at the t-th time period (Figure 8.3):

$$S_i(t)=S_i(t-1)$$

$$+\sum_{c\in R} I_c(t)+\sum_{k\in R} q_k(t)+\sum_{p\in R} x_p(t)-\sum_{h\in R} x_h(t)-\sum_{r\in R} x_r(t)-\delta_i(t) \quad (8.5)$$

$$\forall i,t;R=1,2,....$$

where $S_i(t)$ = state of the i-th reservoir in the t-th time period, with $S_i(0)$ as the initial state at $t = t_0$; $I_c(t)$ = c-th main uncontrolled input of the i-th reservoir at t-th period; $q_k(t)$ = k-th additional uncontrolled input; $x_p(t)$ = p-th main controlled inflows; $x_h(t)$ = h-th controlled output for allocating the amount of water x to user h in period t; $x_r(t)$ = r-th additional controlled output of the i-th reservoir in period t; and $\delta_i(t)$ = minimum admissible amount of water in the i-th reservoir in t-th time period. It is noteworthy that, invariably, the number of questions to be addressed and the uncertainty of the future streamflows, etc. preclude the use of a single mathematical model for river basin planning.

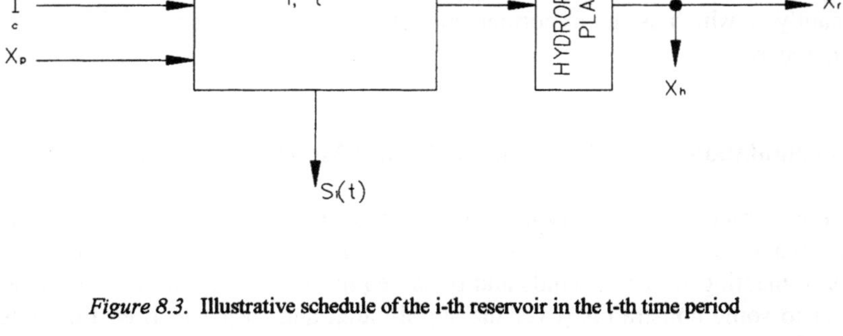

Figure 8.3. Illustrative schedule of the i-th reservoir in the t-th time period

As an alternative path of analysis, the basic configuration of conveyance links, river channels, and reservoirs can be represented as a network. In such a network, the real system's physical elements are represented as a combination of two possible network components – *nodes* and *links/arcs*. As the nomenclature implies, a *node* is a connection or branching point within the network, or both. Therefore, a node is analogous to either a reservoir or a non-storage junction in the physical system. Additionally, a node is a network component that is considered to have the capacity to store a finite and bounded amount of material moving in the network. Reservoirs are represented by nodes with storage capacity – as well as have the ability to serve as branching points. Demands placed on the system are also located at nodal points. Furthermore, any water entering the system, such as might occur naturally from runoff or artificially through imports, is to be introduced at a nodal point. *Arcs* connect all nodes; the transfer of water among the various network nodes is therefore accomplished via links/arcs. Typically, a link is a river reach – or could also be canals or pipes – with a specified direction of flow and fixed maximum and minimum capacities. A nodal equation formulation of our network analyses would be based on an approach that relates the variables describing the individual elements by a set of mass balance equations at each of the nodes, analogous to the continuity of flow in electrical networks. Associated with each junction, *i*, the individual time-varying parameters are interrelated by the following set of nodal mass balance equations:

$$Q_i + \sum_j q_{ij} - R_i - L_i - B_i = 0 \quad i = 1,2,....n \tag{8.6}$$

where Q_i = natural monthly mean discharge in the i-th section; q_{ij} = inflows from upstream *j* reservoirs to *i*, and outflows to downstream *j* reservoirs from node *i* (with the summation being taken over all the nodes connected to node *i*); R_i = releases for monthly average water demands of the consumers 'connected' to section *i*; L_i = evaporation and seepage losses occurring at node *i* (which are a function of the storage volumes); and B_i = net balance at node *i*. The values Q_i, R_i, L_i, B_i represent the i-th elements of *k* vectors having zero elements throughout, except where a reservoir or user appears.

In general, the optimal design of the individual elements in a system of reservoirs acting jointly cannot be assessed in isolation or independently – because the downstream reservoir(s) will receive rather regulated flows with less variability compared with the virgin flows. Furthermore, the structures at the reservoir sites are mutually interrelated, and this situation may influence the design and operation quite significantly – which is an important enough a reason to consider all the reservoirs simultaneously.

8.3. Formulation of the Reservoir Model and Modeling Methodology

The overall objective of the typical reservoir design model is to find the optimum storage capacities of a reservoir system within a given river basin that satisfies certain policies of meeting water demands and other requirements at minimum investment cost – subject to some constraints (over the supply area and over specified time intervals); this would typically entail a benefit-cost analysis of the problem with fixed benefits of meeting all demands. Certain simplifying assumptions are usually incorporated into the modeling schedule in order to produce a model that can be implemented with reasonable level of effort.

Invariably, a decomposition of the system into subsystems becomes necessary in order to allow for a more feasible and realistic analysis of complex reservoirs systems. It is assumed that the system derives its supply from the randomly distributed natural inflows. Further to this, regulated inflow from upstream reservoir releases, diversion return flows, etc., may arrive as inputs to any reservoir. In fact, the inflow into any branch of the system during any time period cannot be analyzed in isolation from the other branches, or from inflows in other time periods. Considering the fact that the explicit appraisal of the multivariate inflow process tends to pose substantial difficulties in the case of multireservoir analyses, the formulation presented here utilizes an approach comprising of the development of a mathematical model of the multivariate river flow process – which is used to generate synthetic inflow traces for the system; subsequently, the system is simulated over these traces. A further assumption is that the operational rules are fixed *a priori*; these rules are taken to be same for all states of the system. This implies that the outputs are random but deterministic (i.e., definite) demands, and so are evaporation and seepage losses as well as pumpage drafts, etc.

In the multireservoir system configuration, each reservoir could be connected to every other reservoir and each could receive discharges from any or all other reservoirs, as dictated by the particular system configuration; furthermore, the reservoirs can be acting in tandem or be connected in parallel. The multiple reservoir system can then be modeled on the state equations of the reservoirs, according to the continuity principle – and a mathematical model can be formulated on this basis. For a typical multireservoir interconnection (as depicted in Figure 8.4), the mass conservation equation for any one reservoir *i* in the system of *n* reservoirs during the time period *t* is (Figure 8.5):

$$S_i(t) = S_i(t-1) + Q_i(t) - R_i(t) - \varepsilon_i S_i(t) - L_i(t) \tag{8.7}$$

where $S_i(t)$ = storage in i-th reservoir at time t [Mm^3]; $Q_i(t)$ = total inflow into the i-th reservoir in time [m^3s^{-1}], the components of which are discussed below; $R_i(t)$ = overall 'instantaneous' release from i-th reservoir in time *t* [m^3s^{-1}], the details of which are given further elaboration below; $\varepsilon_i S_i(t)$ = evaporation loss from the i-th reservoir, as a fraction of storage, at time *t* [m^3s^{-1}]; $L_i(t)$ = other losses from i-th reservoir, such as seepage losses, in time *t* [m^3s^{-1}]; and T = end of the simulation horizon, for a time domain of $t \in [0, t]$. The storage in each reservoir is further constrained between a lower/minimum limit, $S_{i,min}$, and a maximum/upper limit, $S_{i,max}$, viz.:

$$S_{i,min} < S_i(t) < S_{i,max} \qquad i = 1,2 \ldots\ldots n; \quad t \in [0, T] \tag{8.8}$$

In general, the reservoir total inflow $Q_i(t)$ is made up of some or all of several possible components – most prominently, the following:

(i) Natural inflow, represented as a stochastic unregulated inflow into reservoir *i* in period $t \in [t-1, t]$, represented by $q_i(t)$ [m^3s^{-1}].
(ii) Inflow from upstream reservoirs – i.e., if j-th reservoirs is upstream of i-th reservoir, then inflow from the upstream j-th reservoir is $U_i(t)$ [m^3s^{-1}].
(iii) Return flows from m-th user serving as inflow to i-th reservoir, $W_{im}(t)$ [m^3s^{-1}]; this would normally comprise irrigation and diversion return flows, and other imports.
(iv) Pumpage return flows from the k-th source (e.g., pumps, reservoirs, etc.) to the i-th reservoir in time *t*, represented by $P_{ik}(t)$ [m^3s^{-1}].

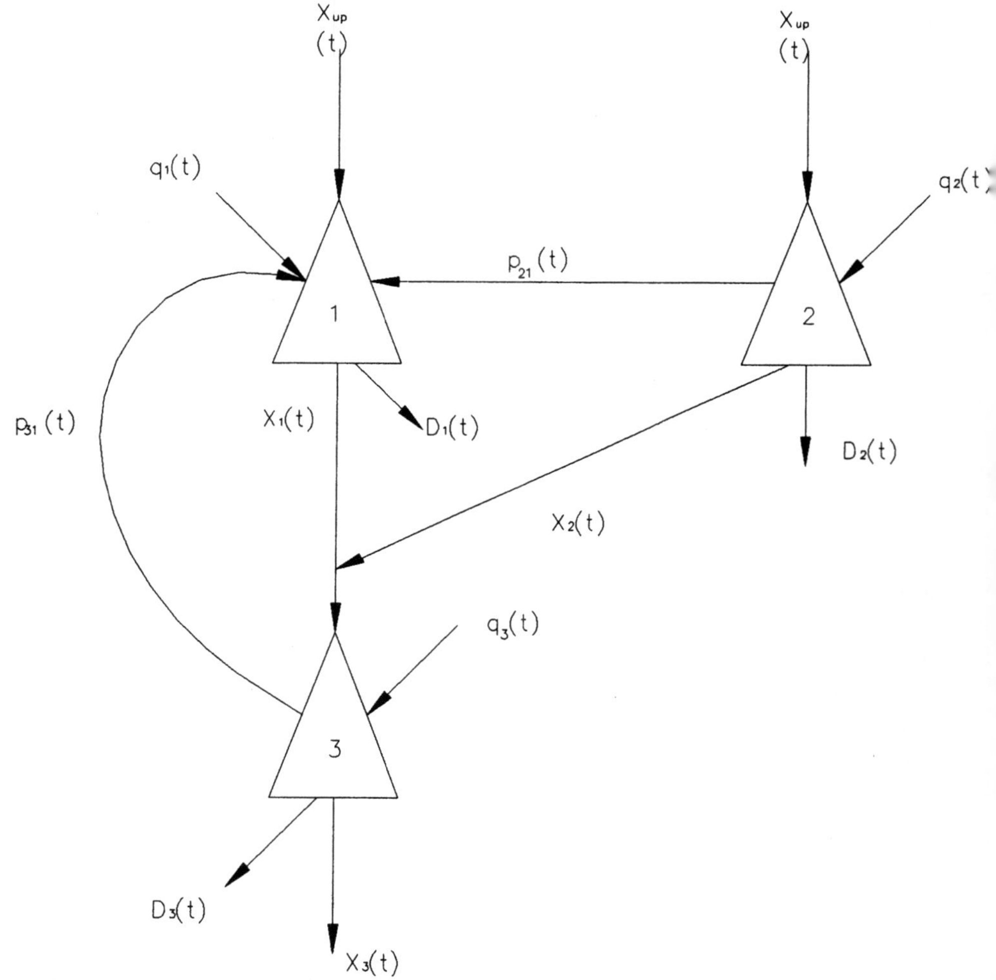

Figure 8.4. **A schematic of typical interconnection for a multireservoir system**

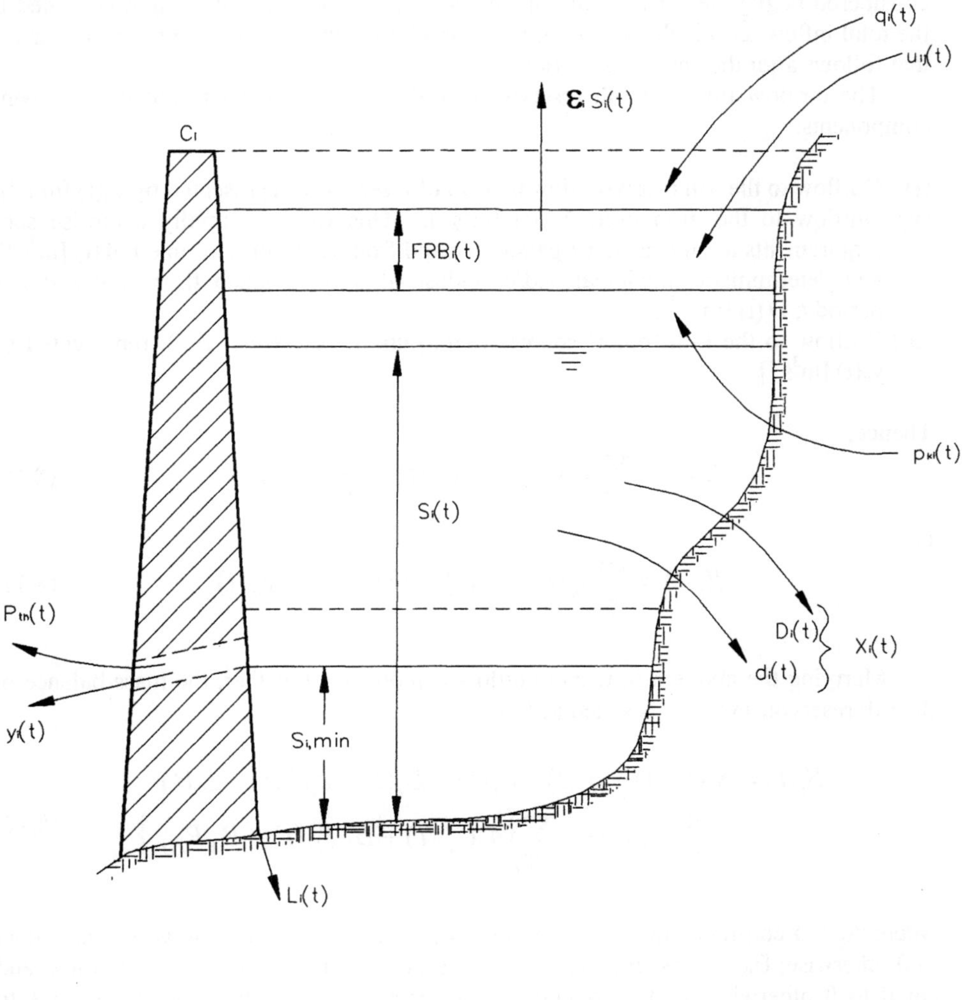

Figure 8.5. **An isolated unit from a linked-up reservoir system**

Consequently,

$$Q_i(t) = q_i(t) + \sum_j U_{ij}(t) + \sum_m W_{im}(t) + \sum_k P_{ik}(t) \tag{8.9}$$

It should be acknowledged that there are time delays characterizing these inflows from upstream reservoirs as well as the return flows, etc.; these time delays can however be considered negligible for ease of computation. A further item that might be added to the total inflow term is that resulting from direct precipitation over the reservoir surface that follows after the dam construction.

The outflow from the i-th reservoir can also be broken down into the following components:

(i) Outflow to the j-th reservoir downstream of reservoir *i*, represented by $v_{ij}(t)$ [m^3s^{-1}].
(ii) Outflow to the m-th user, $X_{im}(t)$ [m^3s^{-1}]. This would normally comprise such requirements as stochastic irrigation demand from reservoir *i* at time *t*, $d_i(t)$ [m^3s^{-1}]; and deterministic municipal and/or industrial water demand from reservoir *i* in period *t*, $D_i(t)$ [m^3s^{-1}].
(iii) Outflow to the k-th 'other' sources (e.g., pumps, reservoirs, etc.), represented by $y_{il}(t)$ [m^3s^{-1}].

Thence,

$$R_i(t) = \sum_j v_{ij}(t) + \sum_m x_{im}(t) + \sum_k y_{il}(t) \tag{8.10}$$

or

$$R_i(t) = \sum_j v_{ij}(t) + d_i(t) + D_i(t) + \sum_k y_{il}(t) \tag{8.11}$$

Marrying the above inflow and outflow components together, the mass balance of the i-th reservoir in time *t* is stated as follows:

$$\begin{aligned} S_i(t) = {} & S_i(t-1) + q_i(t) - d_i(t) - D_i(t) - \varepsilon_i S_i(t) - L_i(t) \\ & - \sum_l y_{il}(t) + \sum_{k=1,k\neq i}^{n} \left\{ A_{ij} U_{ij}(t) + B_{ki} P_{ki}(t) - B_{ik} P_{ik(t)} \right\} \end{aligned} \tag{8.12}$$

where A_{ij} = a constant, equal to 1 if reservoir *j* releases flow into reservoir *i*, and equal to 0 otherwise; B_{ik} = constant, equal to 1 if reservoir *i* pumps/diverts to reservoir *k*, and equal to 0 otherwise; $P_{ki}(t)$ = quantity of water pumped/diverted from reservoir *k* to reservoir *i* in period *t*; and $P_{ik}(t)$ = quantity of water pumped/diverted from reservoir *i* to reservoir *k* in period *t*. Based on this formulation, the design variables can be identified and the evaluation and synthesis of the design model carried out subsequently.

8.3.1. USING MONTE CARLO TECHNIQUES IN THE OPTIMAL DESIGN OF RESERVOIR SYSTEMS

The *Monte Carlo* method is concerned with mathematical (or conceptual) experiments involving random numbers (Hammersley and Handscomb, 1965). By using Monte Carlo techniques to generate long or multiple streamflow records, and then analyzing the output of a water resource system with these inputs, it is possible to test the system's response to periods which are longer and which contain greater extreme events than does the historical record. In addition, we can obtain an estimate of the probability of failure associated with a given design. Indeed, the basic simplicity of Monte Carlo analysis permits its use in the study of the most complex reservoir systems without any alteration being necessary in the fundamental approach. Its accuracy is dependent on the accuracy of the mathematical model of the system and the ability of the generating technique to simulate representative future streamflow records. Fundamentally, the Monte Carlo method of analysis can be said to be a stochastic implicit approach of design – since it belongs to the class of implicitly stochastic models.

The basic requirement in a Monte Carlo analysis of a reservoir system is a deterministic mathematical model that simulates the response of the reservoir system. The input to the reservoirs is a set of generated streamflow records equal in length to the design period, and which are identical to the historical record with regards to those characteristics that are of significance for reservoir design. The initial conditions are re-established at the start of each record, and a note is made of whether or not the system fails to meet its pre-set demands at all points in the record. The fraction of generated records during which the system fails to meet the project goals is then defined to be the probability of failure under the given set of conditions.

It is noteworthy that the inflow to reservoirs are stochastic functions, with very high variability common to those rivers where firm-water supplies are most essential to economic development in a given geographical locality. This characteristic of stochasticity is incorporated in a design procedure by way of Monte Carlo analyses – among other methods, such as critical-period analyses and expected-value optimization; in fact, these methodologies tend to complement each other, and as such are not necessarily independent. It is also noteworthy that, to analyze and design a reservoir system on the assumption that some form of variable release rule is to be used has the advantage of simulating the system more closely to the manner in which the system will be operated when in use.

8.4. A Multisite Multiseason Flow Generation Strategy

When dealing with a situation involving several sites, it is not a satisfactory approach to simply use single site generation procedures for each of the sites in turn – since flows at the various sites can be strongly correlated. Independent generation of the flow values for multiple sites cannot preserve spatial and temporal correlation between flows – and thus the need for multivariate techniques in multisite basin simulations. Such multivariate models tend to preserve not only the serial correlational (i.e., temporal) structure for each variable, but also the cross-correlational (i.e., spatial) structure between all pairs of variables. For n sites for which synthetic streamflows are desired, it very often is necessary to consider the variations of flows within each year –

streamflows in most areas having within-year variations (e.g., exhibiting wet and dry periods); a multiseason multisite approach which generates within-year flows is therefore adopted under such circumstances – with the sample data set being assumed to be one of homogeneous series.

The *disaggregation model* (also referred to as the *method of fragments*), which allows for the generation of synthetic flows that reproduce the relevant statistics both at the *annual level* and at the *seasonal level*, is often to be employed in multiseason multisite studies; annual flows for the *n* sites in question form the input to this disaggregation model. Sequences of annual flows that preserve the relevant statistical properties (i.e., observed means, standard deviations, lag-0 and lag-l correlations and cross-correlations, etc.) of the annual data are generated – for example, by using the Matalas multivariate first-order autoregressive annual model (i.e., a multivariate stationary lag-one Markov model); these annual flows are subsequently disaggregated to monthly (or other appropriate seasonal) values. It is noteworthy that, it is important to generate sequences equal in length to the period of interest (usually equal to the project life span) – recognizing that one long sequence broken into subsequences yields interrelated sequences unsuitable for making probabilistic statements about the expected system performance.

A brief outline of the mathematical basis for the multisite and multiseason models is enumerated below. More detailed discussions can be found elsewhere in the hydrological literature (e.g., Linsley *et al.*, 1982; Loucks *et al.*, 1981; Salas *et al.*, 1980; Kindler and Zuberek, 1967; Clark, 1973; Valencia and Schaake, 1973; Fiering and Jackson, 1971; Matalas, 1967; and Fiering, 1967).

8.4.1. THE MULTIVARIATE AUTOREGRESSIVE AR(l) MODEL FOR THE MULTISITE ANNUAL GENERATION SCHEME

The annual streamflow values that form the input to the disaggregation model (see Section 8.4.2) can be generated by the multivariate first-order autoregressive sequential annual model – which takes the following general form (assuming a zero-mean and unit-variance notation, with variables at each of *n* sites that have been transformed to multivariate normal):

$$Z_t = AZ_{t-1} + B\varepsilon_t \tag{8.13}$$

where Z_t is a *n*-dimensional (i.e., *n x 1*) vector of values assumed by *n* hydrological variables (e.g., annual streamflow values for *n* sites) at time (year) *t*, so that:

$$\{Z_t\} = \begin{bmatrix} Z_{1t} \\ Z_{2t} \\ . \\ . \\ Z_{nt} \end{bmatrix} \tag{8.14}$$

ε_t is a *n*-dimensional vector of normally- and independently-distributed random variables with zero mean and unit variance, so that: $\varepsilon_t \sim N(0,1)$. The random vector ε_t is assumed to be uncorrelated in time and space, i.e., $E[\varepsilon_{it}\,\varepsilon_{jt}] = 0$ for $i \neq j$, or $E[\varepsilon_t\,\varepsilon_t^T] = I$; I is the identity matrix and T denotes the transpose of the matrix, with E being the expectation notation. *A*, *B* are (*n x n*) square matrices of parameters to be estimated from the *n* data sequences, the first of which is:

$$\{Z_{11}, Z_{12}, Z_{13}\ldots\ldots Z_{1n}\} \tag{8.15}$$

and where such sequence is assumed stationary, with each sequence containing *n* elements; they specify the form of time and space interdependence of flows. The matrices *A* and *B* are estimated by the following expressions:

$$A = M_1 M_0^{-1} \tag{8.16}$$

$$BB^T = M_0 - M_1 M_0^{-1} M_1^T \tag{8.17}$$

where:

M_0 = matrix of zero lag cross covariances (i.e., correlation matrix)

$= E[Z_t Z_t^T] = E[Z_{t-1} Z_{t-1}^T]$, assuming weak stationarity

$$= \begin{bmatrix} \frac{1}{N}\sum Z_{1t}Z_{1t} & \frac{1}{N}\sum Z_{1t}Z_{2t} & \cdot\ \cdot & \frac{1}{N}\sum Z_{1t}Z_{nt} \\ \frac{1}{N}\sum Z_{2t}Z_{1t} & \frac{1}{N}\sum Z_{2t}Z_{2t} & \cdot\ \cdot & \frac{1}{N}\sum Z_{2t}Z_{nt} \\ \cdot & \cdot & \cdot\ \cdot & \cdot \\ \cdot & \cdot & \cdot\ \cdot & \cdot \\ \frac{1}{N}\sum Z_{nt}Z_{1t} & \frac{1}{N}\sum Z_{nt}Z_{2t} & \cdot\ \cdot & \frac{1}{N}\sum Z_{nt}Z_{nt} \end{bmatrix}$$

M_1 = matrix of lag-one cross covariances (i.e., correlation matrix) $= E[X_t X_{t-1}^T]$

$$= \begin{bmatrix} \frac{1}{N}\sum Z_{1,t}Z_{1,t-1} & \frac{1}{N}\sum Z_{1,t}Z_{2,t-1} & \cdot\ \cdot & \frac{1}{N}\sum Z_{1,t}Z_{n,t-1} \\ \frac{1}{N}\sum Z_{2,t}Z_{1,t-1} & \frac{1}{N}\sum Z_{2,t}Z_{2,t-1} & \cdot\ \cdot & \frac{1}{N}\sum Z_{2,t}Z_{n,t-1} \\ \cdot & \cdot & \cdot\ \cdot & \cdot \\ \cdot & \cdot & \cdot\ \cdot & \cdot \\ \frac{1}{N}\sum Z_{n,t}Z_{1,t-1} & \frac{1}{N}\sum Z_{n,t}Z_{2,t-1} & \cdot\ \cdot & \frac{1}{N}\sum Z_{n,t}Z_{n,t-1} \end{bmatrix}$$

The matrix BB^T is necessarily symmetric, and B can be assumed to be lower triangular of the following form without loss of generality (Young, 1968):

$$B = \begin{bmatrix} b_{11} & 0 & 0 & \cdot & \cdot & 0 \\ b_{21} & b_{22} & 0 & \cdot & \cdot & 0 \\ b_{31} & b_{32} & b_{33} & \cdot & \cdot & 0 \\ \cdot & \cdot & \cdot & \cdot & \cdot & \cdot \\ \cdot & \cdot & \cdot & \cdot & \cdot & \cdot \\ b_{n1} & b_{n2} & b_{n3} & \cdot & \cdot & b_{nn} \end{bmatrix} \tag{8.18}$$

This then forms the basis for determination of the B matrix. Alternatively, application of the principal component technique that seems more general may be employed to determine matrix B. Indeed, comparative computations (Kindler and Zuberek, 1976) indicate no superiority of any one of the alternative techniques.

Normalization of the Historical Flow Data

The Matalas algorithm expects normally distributed input flows, and if highly skewed flows are used, it is to be expected that negative simulated flows would be generated. This can, somehow, be circumvented by an optional transformation of the historical flow data – with a subsequent convolution back into the untransformed sample space. Thus, if data used have been 'normalized' by some one transformation or another (e.g., by standardizing the flow records, or working with the logarithm of flows, etc.) – as may be necessary to validate the model – then the final/actual flows are ultimately obtained by an inverse transformation (i.e., a 'de-normalization') to that used earlier.

As hydrologic data is rarely multivariate normal, the optional transformations can be applied in an attempt to make the application of the algorithm more nearly theoretically correct. However, the use of this transformation does not guarantee that the resulting sequences will be multivariate normal, and deviations can lead to anomalies and biases in the expected results (Ciriani *et al.*, 1977). Some simple normalizing transformation options that are typically incorporated into the flow generation procedure are provided below – with accompanying standardization by a linear model given by:

$$Z = \frac{(y-u)}{\sigma} \tag{8.19}$$

- NATURAL LOGARITHMS – i.e., use of the logarithm of flows instead of the raw flow data, so that:

$$y = \log(x) \ and \ Z = \frac{\ell n(x) - \mu_y}{\sigma_y} \tag{8.20}$$

with a corresponding inverse transformation of:

$$x = \exp(y) = \exp(\mu_y + \sigma_y z) \tag{8.21}$$

- POWER TRANSFORMATION – SQUARE ROOT, such that:

$$y = \sqrt{x} \ \ and \ \ Z = \frac{y - \mu_y}{\sigma_y} \tag{8.22}$$

with the corresponding inverse transformation of:

$$x = y^2 \text{, where } y = (\mu_y + \sigma_y z) \tag{8.23}$$

- POWER TRANSFORMATION – CUBE ROOT, so that we have the standardized variates:

$$y = \sqrt[3]{x} \ \ and \ \ Z = \frac{y - \mu_y}{\sigma_y} \tag{8.24}$$

with the corresponding inverse transformation of:

$$x = y^3 = (\mu_y + \sigma_y z)^3 \tag{8.25}$$

Indeed various other normalizing transformations can be used, but it is always more convenient to adhere to the simple ones – as long as they satisfy the design requirements. A qualitative criteria for selection of the transformation that brings the natural flows closest to the normal distribution is based on a comparison of the values of Skewness (i.e., Skewness→0 for normal distribution) and Kurtosis (i.e., Kurtosis→3 for normally-distributed variates) estimated for the sequences that have been 'normalized' by application of the different transformations procedures. The final selection of the appropriate one is then left to the analyst's judgment.

Estimation of the Sample Parameters and Model Fitting

For each of our *n* potential sites, the observations in the time period *t* would be

$$x_{it} = \{x_{i1}, x_{i2}, \ldots\ldots\ldots\ldots x_{it}\},\ i = 1,2,\ldots\ldots n \tag{8.26}$$

These should represent a normally distributed data set, otherwise they are so-transformed using the appropriate transformation to render them as such – so that we obtain the normal or normalized sequence given by:

$$y_{it} = \{y_{i1}, y_{i2}, \ldots\ldots\ldots\ldots y_{it}\},\ i = 1,2,\ldots\ldots n \tag{8.27}$$

Next, we standardize the sequences, to render them stationary (or near-stationary), by using the following linear model:

$$Z_{it} = \frac{y_{it} - \hat{\mu}_t}{\hat{\sigma}_y} \tag{8.28}$$

The means that, as appropriate, standard deviations and correlation coefficients that are needed for the determination of the lag-0 and lag-l correlation matrices of Z_t can now be estimated – after which the AR(1) model for the z_t series is fitted in accordance with equation (8.13), i.e., $Z_t = AZ_{t-1} + B\varepsilon_t$.

8.4.2. THE DISAGGREGATION OF ANNUAL STREAMFLOW DATA

The priori generated series obtained in Section 8.4.1 is now input to the disaggregation model along with a series of random normal deviates; these annual flows is then disaggregated to seasonal values upon input into the disaggregation model that takes the following form:

$$X = GZ + HV \tag{8.29}$$

where:

X = vector of nT transformed normally distributed seasonal flows, $x_{i,t,\tau}$, in season t at site i in year τ = $(x_{11\tau}, \ldots\ldots x_{1t\tau}, \ldots x_{2t\tau}, \ldots\ldots x_{n1\tau}, \ldots\ldots x_{nt\tau})^T$

Z = column vector of n annual transformed (standardized normally-distributed) flows = $(z_1, \ldots\ldots\ldots z_n)^T$ and $z_i = \sum_{t=1}^{t} x_{it}$

V = a $(nt \times 1)$ vector of random numbers drawn from N(0,1)

G, H = coefficient matrices of dimension $(nt \times n)$ and $(nt \times nt)$ respectively, so-selected as to reproduce the observed correlations among the elements of x_τ and between the elements of x_τ and z_τ.

The model parameter estimates are obtained based on the following relationships:

- $S_{ZZ} = E[ZZ^T] = \{1/(n-1)\}\Sigma[ZZ^T]$, i.e., the matrix of covariances among the annual series.
- $S_{XX} = E[XX^T]$, i.e., the matrix of covariances among the seasonal series.
- $S_{XZ} = E[XZ^T]$ and $S_{ZX} = E[ZX^T]$, i.e., the matrices of covariances between the series and the annual series.

Thence,

$$G = S_{XZ} S_{ZZ}^{-1} \tag{8.30}$$

$$HH^T = S_{XX} - GS_{ZX} = S_{XX} - S_{XZ} - S_{XZ} S_{ZZ}^{-1} S_{ZX} \tag{8.31}$$

Decomposition of HH^T to obtain a value of H is effected via principal component analysis – using the properties of eigenvalues and vectors – otherwise, the lower triangular matrix notation is used; the model is then fitted by stage-wise disaggregation. The disaggregation model produces sub-series with zero mean, unit variance – and in a normalized form; the appropriate inverse transformation should therefore be applied to obtain the true generated sequences. A noteworthy observation is that, the disaggregation model apparently does not conserve mass properly due to the limited portions of the correlation matrices built into the model structure (Linsley *et al.*, 1982). An optional correction factor may therefore be applied for each site to ensure that the sum of the seasonal flow volumes equals the annual total; Linsley *et al.* (1982) suggest a simple adjustment scheme, as follows:

$$Q_{\tau,t} = \frac{\sigma_t}{\sum_{t=1}^{t} \sigma_t} \Delta\tau + x_{\tau,t} \tag{8.32}$$

where $Q_{\tau,t}$ = corrected flow; σ_t = standard deviation of the flow in season *t*; and $\Delta\tau$ = difference between the annual flow and the sum of the seasonal flows.

8.4.3. IMPLEMENTATION OF THE FLOW GENERATION ALGORITHM

The overall scheme for the flow generating procedure is comprised of the parameter estimation and model fitting of the multisite multiseason model. The process utilizes the AR(l) (Section 8.4.1) and the disaggregation (Section 8.4.2) models. Specifically, the flow generation process for the synthetic flows is a disaggregation process in which the historical monthly flows are aggregated into annual flows, and then the AR(l) model is used to generate a set of synthetic sequences of annual events considered long enough to equal the project life. Subsequently, the annual events generated can be disaggregated stagewise and sequentially – first into quarterly and then monthly flows.

At each level of aggregation, mean values, standard deviations, skewness, kurtosis and all lag-0 and lag-l correlation coefficients are estimated and compared. In a detailed and/or sophisticated study, the statistical resemblance between the historical and synthetic streamflow sequences may be evaluated using methods such as the Student's t-test (mean values), Z-Fisher test (variances), and log-transform test (correlation coefficients).

8.5. Storage-Capacity Allocation to Reservoir Sites

The overall reservoir capacity, CAP_i, comprises the active storage (for water supply and flow regulation), the dead storage (for sediment, recreation and hydropower), and the flood control storage (required in the flood-prone seasons). In general, we should design our system according to the constrained objective of minimizing the overall storage capacity at each site *i*, viz.:

$$Minimize\, CAP_i \begin{cases} \geq V_i^{dead} + V_i^{active} + V_i^{flood}, \forall t \text{ in flood seasons} \\ \geq V_i^{dead} + V_i^{active}, \forall t \text{ in non-flood season} \end{cases} \tag{8.33}$$

Typically, flood storage need not be fixed – because flood control space can be managed by setting maximum allowable storage levels during months with flood risk. On the other hand, once established for a design, the dead storage capacity is fixed for the life of the structure – especially because, among things, it is an essential factor in the design and placement of the power plant, the turbines, and other outlet works for hydropower requirements. In considering the active storage component, typically firm yield analyses would first be conducted for various sizes of carryover storage to determine the maximum firm yield that could be delivered from each size of reservoir –

taking into account reservoir seepage, evaporation, and the variable volumes of water to be placed in storage in the reservoir during each time interval.

The evaluation of complex reservoir systems involves so many issues – albeit, our focus here is on the allocation of storages to multiple sites, and for multiple storage space requirements. With the general configuration of the system defined, the set of alternative water use and control measures specified, and the hydrologic data organized in its utilizable form, the next concern is to establish the physical limits of the system components – namely, the range of storage capacity at the various reservoir sites. Assuming that all other non-hydrological factors (e.g., accessibility, cost of conveyances, etc.) are proportionately equal for selected users, weighting terms/factors – hereafter called *sizing factors* – can be utilized in the reservoir capacity estimation process; these so-called sizing factors are used to designate the different joint policies that could be simulated for the system under consideration.

8.5.1. DETERMINATION OF THE RESERVOIR SIZING FACTORS

The matrix of sizing factors – necessary to simulate different reservoir planning and/or operating policies – is defined as follows:

$$A_r = \{a_{ijt}\}, \quad r \in \mathrm{Re}\,al\ Integer\,(1{,}2.......) \tag{8.34}$$

The coefficients/elements of this matrix may be assumed arbitrarily and varied at the discretion of the decision-maker to reflect on possible exclusion, or otherwise, of some specific sites to certain users. Thus, A_r may be established for each specific reservoir use (such as for water supply, irrigation, etc.), or the same A_r may be assumed for all uses. In any case, the elements, a_{ijt}, of matrix A_r should satisfy the following constraints:

$$0 \leq a_{ijt} \leq 1 \tag{8.35}$$

That is, in the t-th month, the i-th reservoir supplies a fraction, (a_{ijt} x B_{jt}), of the water deficit, B_{jt}, of the j-th user's requirements. Furthermore, this condition should be met for each t-th month and for any j-th user – so that:

$$\sum_{i=1}^{n} a_{ijt} = 1 \tag{8.36}$$

Indeed this has the implicit assumption that the runoff regulation system is capable of theoretically eliminating all water shortages in the basin. An alternative formulation of this constraint – which attaches some probability of failure and/or reliability statement to it – would be to state it as follows: $\Sigma a^{*}_{ijt} \leq 1$, where again $0 \leq a^{*}_{ijt} \leq 1$.

In order to establish a consistent policy, it is proposed that the a_{ijt} elements of the A_r matrices be derived according to the following relation:

$$a_{ijt} = \frac{Q_{it}}{\sum_{i=1}^{n} Q_{it}} \tag{8.37}$$

where Q_{it} represents the monthly discharge value for site i in month t. The concomitant $[n \times 12\tau]$ matrix, A_r, that reflects the different design policies, is thence determined for each trace of length τ years for the synthetic flow obtained earlier in Section 8.4. Typically, for a simulation time of τ (years) = 12τ months ≡ T^* months and n sites,

$$A_r = \begin{bmatrix} a_{1j1} & a_{1j2} & \cdot & a_{1jt} & \cdot & a_{1jT^*} \\ a_{2j1} & a_{2j2} & \cdot & a_{2jt} & \cdot & a_{2jT^*} \\ \cdot & \cdot & \cdot & \cdot & \cdot & \\ a_{ij1} & a_{ij2} & \cdot & a_{ijt} & \cdot & a_{ijT^*} \\ \cdot & \cdot & \cdot & \cdot & \cdot & \cdot \\ a_{nj1} & a_{nj2} & \cdot & a_{njt} & \cdot & a_{njT^*} \end{bmatrix} \quad (8.38)$$

For any one use, this sizing factor forms the basis for the computation of storage requirements. It is noteworthy that, although a monthly time-base is utilized in this example discussion, the systems analyst or designer may opt for a different time-step without affecting the underlying principles for this procedure.

8.5.2. INCORPORATING A RELIABILITY MEASURE

The notion of having to meet all target demand as noted above implies a 100% system reliability, and therefore no failures and/or water shortages under whatsoever. This certainly would be a rather optimistic goal, and could indeed result in an uneconomical over design. It is proposed here, therefore, to include a multiplicative factor, α, which would attach reliability (or failure probability) values to the estimated storages. Thus, with the modified constraint,

$$\sum_{i=1}^{n} a_{ijt} \leq 1 \quad (8.39)$$

a modification can be made to the derivation of the a_{ijt} elements of the A_r matrices, so that we now have:

$$a^*_{ijt} = \alpha \frac{Q_{it}}{\sum_{i=1}^{n} Q_{it}} \quad (8.40)$$

or

$$\sum_{i=1}^{n} a^*_{ijt} = \alpha \sum_{i=1}^{n} a_{ijt} \leq 1 \quad (8.41)$$

where α, is a measure of reliability. Thence, the necessary reservoir capacities can be estimated for different levels of reliabilities; this is illustrated in Box 8.1 below. Furthermore, a graphical relationship can be established that could enable us to adjust our design values according to what permissible failure probabilities have been specified for the particular project (see illustration in Figure 8.6). It is noteworthy that, rather than attach the same reliability (α) values for all sites and for all periods of the simulation time, different reliability values (α_{ijt}) may be specified and used for the computations.

Box 8.1. Reservoir capacities for different levels of reliabilities at the potential reservoir sites

α = / ***Site***	**100%**	**99%**	**95%**	**90%**	**80%**	**70%**	.....	.....	α (%)
1	$V_{1,100}$	$V_{1,99}$	$V_{1,95}$	$V_{1,100}$	$V_{1,100}$	$V_{1,100}$	·	·	$V_{1,\alpha}$
2	$V_{2,100}$	$V_{2,99}$	$V_{2,95}$	$V_{2,90}$	$V_{2,80}$	$V_{2,70}$	·	·	$V_{2,\alpha}$
·	·	·	·	·	·	·	·	·	·
·	·	·	·	·	·	·	·	·	·
·	·	·	·	·	·	·	·	·	·
i	$V_{i,100}$	$V_{i,99}$	$V_{i,95}$	$V_{i,90}$	$V_{i,80}$	$V_{i,70}$	·	·	$V_{i,\alpha}$
·	·	·	·	·	·	·	·	·	·
·	·	·	·	·	·	·	·	·	·
·	·	·	·	·	·	·	·	·	·
n	$V_{n,100}$	$V_{n,99}$	$V_{n,95}$	$V_{n,90}$	$V_{n,80}$	$V_{n,70}$	·	·	$V_{n,\alpha}$

8.5.3. MATRIX OF 'DISAGGREGATED' WATER DEMANDS

A matrix of water demands or drafts required of each component reservoir is established from a diagonal matrix of total/lumped drafts (to be satisfied jointly by the reservoir system) and the matrix of sizing factors computed earlier. The diagonal matrix of total drafts is formed from the *a priori* knowledge of water requirements expected from the system (i.e., water required in storage to satisfy the multipurpose uses, such as the amount for water supply plus that needed for irrigation, etc.). In addition, a determination of the proper size of the reservoir must take into account the time variability of demand for the reservoir releases. Thus, for an overall deterministic time-variable demand, R(t), for each month and required from the complete system acting jointly, the 'disaggregated' monthly demands is computed as follows:

$$[A_r]\begin{bmatrix} R_1 & 0 & 0 & \cdot & \cdot & 0 \\ 0 & R_2 & 0 & \cdot & \cdot & 0 \\ 0 & 0 & R_3 & \cdot & \cdot & 0 \\ \cdot & \cdot & \cdot & \cdot & \cdot & \cdot \\ \cdot & \cdot & \cdot & \cdot & \cdot & \cdot \\ 0 & 0 & 0 & \cdot & \cdot & R_{T*} \end{bmatrix} = \begin{bmatrix} R_{11} & R_{12} & R_{13} & \cdot & \cdot & R_{1T*} \\ R_{21} & R_{22} & R_{23} & \cdot & \cdot & R_{2T*} \\ R_{31} & R_{32} & R_{33} & \cdot & \cdot & R_{3T*} \\ \cdot & \cdot & \cdot & \cdot & \cdot & \cdot \\ \cdot & \cdot & \cdot & \cdot & \cdot & \cdot \\ R_{n1} & R_{n2} & R_{n3} & \cdot & \cdot & R_{nT*} \end{bmatrix} \quad (8.42)$$

$$\{[n \times 12\tau] \equiv n \times T^*\} \times \{[12\tau \times 12\tau] \equiv [T^* \times T^*]\} = \{[n \times 12\tau] \equiv [n \times T^*]\}$$

i.e.,

$$[A_r][R(t)] = [R_i(t)] \quad (8.43)$$

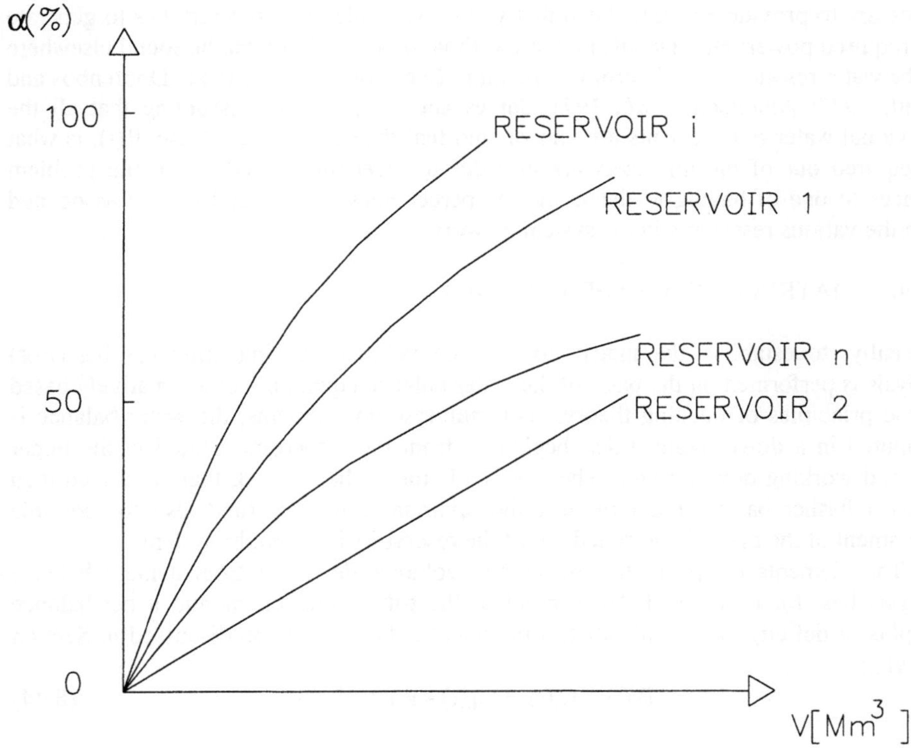

Figure 8.6. **A schematic of cost-efficient capacity [V] vs. reliability [α] relation**

Alternatively, the total monthly (or seasonal) releases can be routed through the various reservoir sites assuming the other sites are nonexistent (i.e., assuming no storages are to be provided at the other sites); the matrix of storage values obtained from this can then be scaled down using the A_r matrix.

Naturally, the components of water demands have to be estimated independently for the region under study. For instance, irrigation water needs should be determined based on factors such as the type(s) of crops and cropping pattern, crop water requirements, and irrigation area; municipal water supply may require population estimates and projections, as well as per capita water demands; industrial water demand would be reflected by the level of industrialization, as well as the projected water demands due to future expansion and new industries; hydropower needs are based storage-area elevation curves – that help determine the necessary amount of storage necessary to provide adequate head that would be available to run turbines to generate the required power; etc. Details on these estimation procedures can be found elsewhere in the water resources and hydrology literature (e.g. Loucks *et al.*, 1981; Doorenbos and Pruitt, 1977; Albertson *et al.*, 1971; James and Lee, 1971). Assuming that all the individual water requirements are known, and that the cumulative release, R(t), is what is required out of the all reservoirs in order to meet these needs, then the problem reduces to one of having to determine the percentages of this total release associated with the various reservoirs in the system network.

8.5.4. MATRIX OF STORAGE VOLUMES

Typically, to establish the matrix for storage volumes, a simulation (or behavior) analysis is performed on the basis of the water balance equation – or alternatively based on the principles of queuing theory. For multireservoir systems, the water balance is computed in a downstream order, beginning from the reservoirs located in the upper part and working downstream. The matrix of storage balance, B, that is derived then forms a further basis for determining the optimum capacities (and also for possible assessment of the operational schedule) of the reservoirs in the multi-system.

The elements of the matrix of storage volumes derivable from a mass balance analysis has, for each i-th balance point in the t-th simulation month, a net balance (surplus or deficit), $B_i(t)$, that satisfies the nodal balance equation discussed in Section 8.2, viz.:

$$B_i(t) = Q_i(t) + \Sigma q_i(t) - R_i(t) - L_i(t) \tag{8.44}$$

From the water balance computations for all sites over the complete simulation time period for the system under study, the distribution in time and space of storage deficits can be deduced. From the [n x T*] matrix of net balance, $B_i(t)$, whose elements are $\Delta S_i(t)$, the [n x T*] matrix of storage depletion, $B_i{}^*(t)$, can be deduced, such that the elements of the latter are defined by:

$$\Delta S_i^*(t) = \sum_{t=1}^{t} \Delta S_i(t) \qquad \forall i,t \tag{8.45}$$

From the [n x 12τ] matrix of storage deficits, $B_i{}^*(t)$, obtained earlier, a [n x 1] matrix of storage capacities, V, is established such that its elements, V_i, are given by:

$$V_i = \max_i \left\{ \left| \min_{i,t} \left\{ \Delta S_i^*(t), 0 \right\} \right| \right\}$$

$$= \max_i \left\{ \left| \min_{i,t} \left\{ \sum^t \Delta S_i(t), 0 \right\} \right| \right\} \tag{8.46}$$

$$= \max_i \left\{ \frac{-1+\lambda}{2} \sum^t \Delta S_i(t) \right\}$$

where: $\lambda = +1$, ...when $\Sigma\Delta S_i(t) > 0$

$= -1$, ...when $\Sigma\Delta S_i(t) \leq 0$

It is noteworthy that, a stage-wise simulation may have to be resorted to when the sizes of the working matrices should become too large and prohibitive.

8.6. Cost-Efficient Capacity Allocations in the Design of Multireservoir Systems

The problem of incommensurability of the different benefits derivable from a water resource project often leads to modeling them separately, which leads to a problem of vector optimization – i.e., instead of one objective function, a whole vector of objectives is optimized without loss of generality. This process gives rise to the so-called concept of *Pareto-optimality* – by which a vector is said to be better than another if it is strictly better in one component and concurrently if it is not worse in all other components.

By using a cost-effective analysis of the system of reservoirs, the storage capacity requirements can be allocated to the various reservoir sites in such a manner as to minimize the cost of storage as a function of the total storage required; a vector of storages is obtained from the different planning policies derived from the different sizing-factor matrices estimated earlier in Section 8.5. The cost of a dam at each dam site can indeed be estimated as a function of the storage capacity provided.

For the models to be capable of identifying cost-efficient reservoir capacities V_i it would be necessary to specify reservoir capacity cost functions, $C_i(V_i)$, for each potential reservoir site *i*. These functions may take various forms – such as a linear cost function (which is but rare, in view of the fact that water resources projects hardly fit into such a function due to the economy of scales of such projects); convex nonlinear functions (which may be gently convex or otherwise); concave nonlinear; etc. In general, the cost functions is represented by the following general type of relationship:

$$C_i(V_i) = [V_i]^\gamma \quad \forall i, \gamma \in \mathrm{Re}al \tag{8.47}$$

Indeed, different cost functional relationships may also be defined for the different potential reservoir sites under investigation (see illustration in Figure 8.7). Typically, these cost functions are discontinuous at $V_i = 0$; also, the cost of a reservoir increases as the height of the dam is increased.

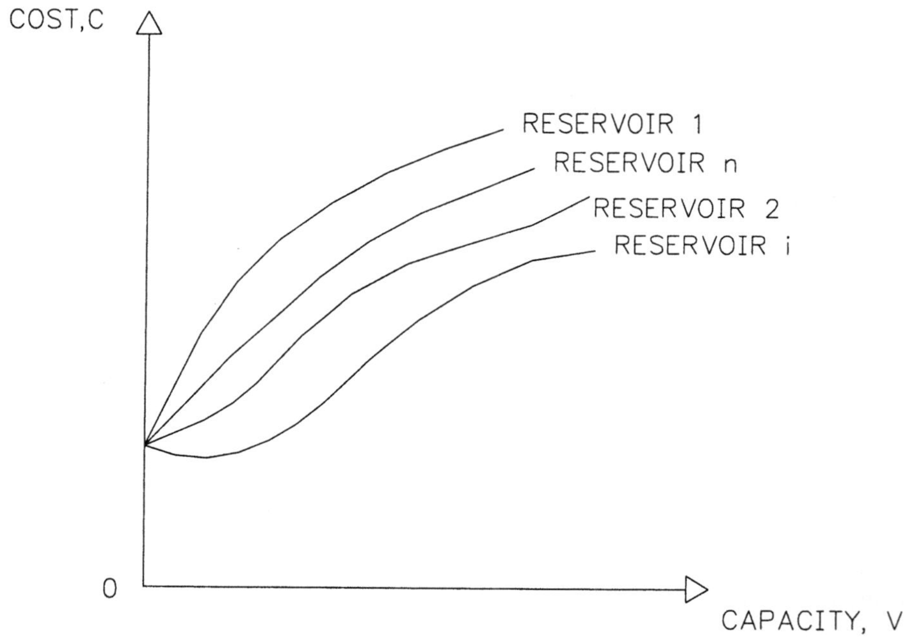

Figure 8.7. **Schematic of cost-capacity relation, showing investment cost functions for various sites**

8.7. An Optimal Solution for the Multireservoir System Design

In the multireservoir design effort, the storage-capacity requirements can be allocated to the various reservoir sites in such a manner as to minimize the total cost of storage as a function of the capacity requirements. To implement the proper solution, the component submodels (discussed in the preceding sections) are married together – its solution being facilitated by use of mathematical models and cost functions.

By using stochastically generated synthetic flow traces, a simulation-optimization process is can be performed deterministically; for each river subsystem, the storage required to meet a given level of output is then allocated to the potential reservoir sites in the most cost-efficient manner. Fundamentally, the simulation procedure starts with a monthly flow regulation algorithm that has monthly inflows, releases, and target outputs. In cases when flood storage is required, provisions should be made for the algorithm to switch to a flood-routing procedure based on a shorter time interval (e.g., one day, or even a few hours interval) when a flood month is encountered; at the end of the flood season, the process can then return to the monthly regulation procedure.

Once the storage spaces, V_i, and investment cost functions, $C_i(V_i)$, of the individual reservoirs are specified, the investment cost of the reservoir system that need to be built (based on different policies, and in turn based on the sizing factor matrix, A_r) can then be estimated. The investment cost of the reservoir system determined based on the matrix A_r is given by:

$$IC_r = \sum_{i=1}^{n} C_i\,(V_i) = \sum_{i=1}^{n} C_i\,[V_i(A_r)] \tag{8.48}$$

In considering different policies – as a consequence of, for instance, working with different A_r (r = 1,2,.........) values – we obtain a different value of investment costs, IC_r (r = 1,2,........), for each policy. The minimum of these IC values can then be considered as the optimum solutions, and the corresponding objective function is:

$$IC_r = IC_r[V_i(A_r)] \rightarrow \min \tag{8.49}$$

That is, the model's objective is to:

$$Minimize\left\{ IC_r = \sum_{i=1}^{n} C_i\,(V_i) \right\} \tag{8.50}$$

Thence, the optimum solution, IC_{min}, corresponds with the storage capacities, $V_{i,opt}$, that need to be designed for the system under study. It has been assumed in this illustrative discussion that the cost of transfers, and operation-maintenance-repair (OMR) costs and the like would remain the same under the different policies. In situation when this is not true, further consideration should be given to this issue – and the appropriate cost components should be included in the overall investment cost that is to be minimized. In effect, the overall objective would be modified as follows:

$$\min\left\{ J = \sum_{i=1}^{n} C_i\,(V_i) + \sum_{i=1}^{n} g(V_i) \right\} \tag{8.51}$$

where $C_i(V_i)$ = investment cost of providing storage capacity, V_i, at reservoir site *i*; and $g(V_i)$ = annual cost to provide spillway outlet capacity, C_i, at reservoir *i*, as well as OMR and other costs. Indeed, OMR may have a very strong effect on minimizing the design costs – because a lower construction cost does not necessarily imply a lower OMR; when an inverse relation can be forecast or is suspected, it might be necessary to opt for a higher investment design capacity. Here again, the analyst may have to use his/her intuition, and/or conduct a more detailed economic benefit-cost/investment-cost studies pertaining to the project.

8.7.1. THE OVERALL MODEL IMPLEMENTATION PROCESS

For river systems of various sizes and geometric configurations, the sequence of steps for the complete implementation of the proposed methodology is provided below.

- *STEP 1 – Streamflow Synthesis.* The multisite generation of seasonal (e.g., monthly) flows up to a specified number of profiles is carried out as a first step (see Section 8.4). As appropriate, generate several sequences of the same length of multisite synthetic data; the length of each sequence is made the same to avoid the possible effect of storage estimates increasing with record length. The length of generated sequence is better made equal to a pre-specified economic life of the project.

- *STEP 2 – Simulation.* This involves routing the generated flow traces in order to estimate the capacities associated with each of the reservoir sites, and for each of the generated sequences. This behavior/simulation analysis is performed in accordance with the algorithm(s) elaborated in Section 8.5. For this analysis to be meaningful, the simulation of each design must be performed with an efficient reservoir system operating policy similar to the one that could, or would be adopted in practice.

- *STEP 3 – Cost-Capacity Analysis.* This comprises establishing the investment costs associated with each of the capacity estimates from each of the generated sequences, and for all the reservoir sites. Based on the V_i values obtained for the alternative policies (in turn, derivable from the A_r matrices), the investment costs of the reservoirs can be read from a previously prepared set of cost function charts that relates capacity to investment costs (as illustrated in Section 8.6). If OMR and conveyance costs are to be given due consideration, then these can also be determined in a similar manner and then incorporated into model. Ultimately, this produces a [*n x r*] matrix of sizing-factor-cum-capacity whose elements are the individual investment costs, as shown in Box 8.2.

- *STEP 4 – Simulation-Optimization.* A simulation analysis is carried out in a deterministic environment, within which an optimization algorithm is nested; this is necessary for the identification of the minimum total investment cost of storage capacity combinations for the reservoir sites. An optimum solution – or at least a 'near-optimal' solution – can be derived if an adequate series of alternative policies pertaining to a wide range of combinations of the A_r matrices is examined. From the [*n x r*] matrix indicated in Step 3 above, the cost summations represented by

$IC_1, IC_2, IC_3, \ldots, IC_r$ are compared – in order to determine the one that produces the minimum overall investment cost, and which indeed is a surrogate to determining the optimal design for a given level of reliability.

This complete process will typically be adapted to implementation with a computer algorithm. Ultimately, the method will help determine the optimal combination of reservoir capacities for a specified level of meeting target releases – and the optimal solution will be the one that results in the lowest cost combination of reservoirs.

Box 8.2. Policy-Capacity matrix

Policy / ***Capacity***	A_1	A_2	A_3	A_4	A_5	.....	.....	A_r
V_1	C_{11}	C_{12}	C_{13}	C_{14}	C_{15}	.	.	C_{1r}
V_2	C_{21}	C_{22}	C_{23}	C_{24}	C_{25}	.	.	C_{2r}
.	.	.	.	.	.	.	.	.
.	.	.	.	.	.	.	.	.
.	.	.	.	.	.	.	.	.
V_i	C_{i1}	C_{i2}	C_{i3}	C_{i4}	C_{i5}	.	.	C_{ir}
.	.	.	.	.	.	.	.	.
.	.	.	.	.	.	.	.	.
.	.	.	.	.	.	.	.	.
V_n	C_{n1}	C_{n2}	C_{n3}	C_{n4}	C_{n5}	.	.	C_{nr}
Σ	IC_1	IC_2	IC_3	IC_4	IC_5	.	.	IC_r

8.8. Optimal Storage Capacity Decisions for Multireservoir Systems

Multiple reservoir decisions arise when we have two or more reservoirs operating for a common purpose; these reservoirs may be acting in series or in parallel. The extent of reservoir interconnection either among the storages themselves or the points of demand determines whether a single or a multireservoir analysis is most appropriate.

In any case, the storages at all potential reservoir sites within a river basin network that are necessary to meet specified drafts in a delineated area depends on the variability of the streamflow, the size of the demand, and the degree of variability of this demand being met. In the complex stochastic problems encountered when designing the storages of a system of several dependent reservoirs, a hybrid model (for the simulation of a multiple reservoir system comprising of the river network, reservoir sites, anticipated water users, etc.) is recommended; thus, a simulation model with optimizing capability may be utilized, such that stochastically generated synthetic data are used for the deterministic simulation. In general, certain simplifying assumptions are usually incorporated in the modeling task – in order to obtain a model that can be implemented with reasonable effort. Ultimately, however, the model of choice should provide an answer to the following basic question: 'what percentage of the total demand required for a supply area should be placed on any one particular reservoir acting in concert with the several other reservoirs in the project area?' Indeed, the primary objective is to develop a model that could define the reservoir capacities required to meet target drafts with specified reliabilities, and at minimum cost.

The so-called sizing factors introduced here for the analysis of the multireservoir systems will allow us to evaluate different policies, and subsequently to select the most attractive design option and/or operational strategy. In fact, the flexibility in the choice of the sizing factors is a great advantage – since it gives the analyst or systems manager a wider telescopic viewing of all apparently feasible designs. Also, a measure of design reliability can be attached to the systems design – which allows the designer to choose a 'safe' and economical value that could result in an optimal design. Furthermore, different levels of reliability can even be employed for the different sites and time periods; this could be very interesting when, for some given reservoir site, time-varying reliabilities are an economically more reasonable basis to utilize in the design effort. Similarly, spatially-varying reliabilities could be employed in the systems design – which is to say that different but acceptable levels of reliabilities can be selected for the various reservoir sites in the system. Indeed, this algorithm can be used to optimize the allocation of capacities to all sites – to meet stipulated demands with desirable reliability.

Chapter 9

HYDROLOGICAL SIZING OF RESERVOIRS FOR FLOOD PROTECTION

A flood wave passing through a reservoir is both delayed and attenuated as it enters and spreads over the pool surface; the process depends on the regime and type of the reservoir. The overall objective in a typical flood protection program is the retention of the water volume of the flood waves, partially or fully in the reservoir – thus enhancing the safety of the downstream areas. From a hydrologic point of view, this means the conversion of the time-function of the upstream discharges into the time-function of the downstream flows. The conversion depends on the geometric and hydraulic features of the pool, as well as on the reservoir operation. Good knowledge of this modifying effect is important to determining the proper design parameters and the reservoir operation schedule.

This chapter introduces some methods of approach that can be employed in the dimensioning of flood protection reservoirs. Essentially there are two main types of the flood protection reservoirs; the first type of the flood control reservoirs are constructed in the river, and the second type belong to the group of emergency reservoirs connected in parallel or in series with the main river. The design and management features of these types of reservoirs are discussed below.

9.1. Determination of the Characteristic Hydrograph

The regulation of flood waves – that involves the lowering of the water level to some specific value – typically can resolve a flood protection problem in the downstream section of a reservoir. However the basic question relating to what happens in the case of extreme flood waves remains. Flood waves that are higher then the characteristic one can usually cause damage at the downstream section of the river (in the absence of a flood control reservoir) or the reservoir (if the allocated flood control storage space is exceeded). Indeed, the extreme flood waves often will endanger the safety of the reservoir and the downstream section of the river. Consequently, there usually is a need for the construction of a special emergency spillway that is capable of safely diverting the extreme floods.

In order to determine the appropriate hydrograph to use in the hydraulic dimensioning of the spillway structures, the maximum measured discharge data of the nearest hydrometric station, NQ, is used to estimate the theoretical probability distribution function, as follows:

$$F(x) = P(NQ \leq x) \tag{9.1}$$

Subsequently, a decision is made on the required safety level for the downstream section of the river. In general, in the case of flood retention reservoirs used for the protection of towns or cities, a 1% probability of exceedance of floods is appropriate; lands in agricultural regions should be protected against a flood flow having 5% to 10 % probability of exceedance. Finally, the selected hydrographs belonging to the flood flows with given probability of occurrence should be presented in a joint system of coordinates, and the average values of ordinates should be standardized using the following ratio:

$$ratio = \frac{NQ(p\%)}{Q^{\max}} \tag{9.2}$$

where NQ(p%) is the flood flow with a p% probability, and Q^{max} is the average value of selected maximum flood flows (Figure 9.1).

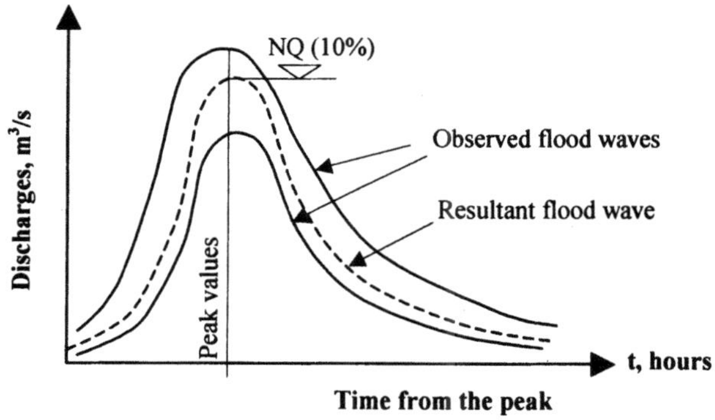

Figure 9.1. Construction of the characteristic flood wave for a probability of 10%

9.2. Spillway and Sluice Gate Considerations in the Design and Operation of Flood Retention Reservoirs

Several issues usually require careful consideration in the construction and operation of flood retention types of reservoirs – amongst which the spillway capacity and sluice gate operations deserve special attention. Indeed, it is possible to manipulate the sluice during the passing of the flood wave; through a continuous regulation of the sluice, there is a possibility for an optimal accommodation of the downstream conditions vis-à-vis the present state of the reservoir – also taking into account the forecasted characteristic discharges. In the case of a fully depleted reservoir, the sluice can be closed. Also, the flood wave can enter into a reservoir having an opened and fixed spillway or an emergency spillway with given fixed sill; the release from this reservoir is dependent on the reservoir depth above the spillway crest – which in turn depends on the inflow and outflow rate. A very important question relates to the process of choosing spillway capacity, and the type of spillway (such as surface, submerged, fuse plug with gates or without gates, etc.). Here, the engineering problem is that the inflow design flood may or may not equal the outflow of the design flood for the spillway capacity. But we also have to take into account the routing and storage effects; the regulating potentials of hydraulic turbines and dam overtopping without failure of breaching; and emergency use of supplemental spillways. Under such types of circumstances, however, it becomes necessary to incorporate a risk analysis of the downstream river section into the procedures used in the determination of the spillway capacity. This means that a comparison should be made of the downstream damages from discharges greater than the spillway capacity *versus* the cost of increasing the design capacity. In this type of analysis, we also have to take into account the possible damages to the spillway, the dam, and/or the power works – which would not necessarily result in failure of the dam. Another aspect of the risk analysis relates to the acceptability of the incremental damages that may occur downstream due to a partial breach or failure of the dam.

Depending on the available hydrological data, solutions to this type of problem may be handled a little differently – mostly dictated by the lengths of hydrologic records. In case of a watercourse that does not have long data series, at least one or more representative flood wave hydrographs should be determined using precipitation data and catchment features. If there is a considerably long data series observed on the watercourse in question (i.e., ≥ 30 years), these data should be utilized to produce the design hydrographs. Having these design hydrographs for the upstream river section of a reservoir, the peak discharge or even the whole shape of the downstream flood wave should be computed. The reservoir space needed for the required attenuation of the flood is also an important outcome of the computational exercise. In the case of an existing reservoir, this attenuation effect depends only on the reservoir operation, whereas in case of a reservoir that is now being designed, the geometric and hydraulic alternatives may also be investigated.

9.2.1. THE CASE FOR A REGULATED SLUICE

The necessary volume of a flood retention reservoir that has a regulated sluice is much lower than the volume of reservoirs constructed with an opened or closed sluice. To enable an accurate determination of the proper flood storage requirements, it is important that there be a hydrometric station on the upstream section of the river – to provide a continuous forecast about the discharges entering into the reservoir. There is also a need for the continuous monitoring of the water level at the downstream section of the river.

The sluice of the reservoir has to be regulated in such a way that, the permissible maximum discharges continue to be lowered up until when the complete emptying of the reservoir occurs. Following that, there is a continuous flow through the opened sluice till such time when the discharge is not higher than the permissible maximum discharges of the downstream section. When the inflow is higher, there is a need to narrow down the sluice to ensure compliance with the permissible discharges; in that case, the water level of the reservoir will increase, and consequently, we have to provide storage for the water volume above the permissible capacity of the downstream river section.

A graphical analysis of the hydrographs can be used to construct the series of curves of the efficiency function of the regulated reservoirs (characterized by probability, p), in accordance with the following relationship:

$$K = f(NQ_d, p) \tag{9.3}$$

Figure 9.2 provides an example of such a hydrograph – presented for a Hungarian stream with a peak discharge of 50 m^3/s and a 1% of probability of occurrence.

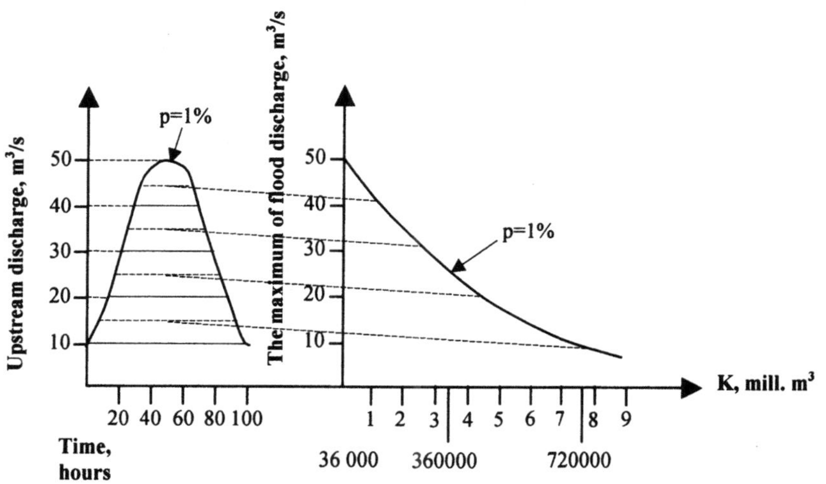

Figure 9.2. Volume of the Volgyseg reservoir

Besides the characteristic hydrograph, the corresponding efficiency function can also be constructed – with the vertical axis in that system of coordinates representing the maximum discharges of the downstream section and the necessary volume of the reservoir on the horizontal axis (Figure 9.3). On the left-hand side of Figure 9.3, the flood hydrographs of the investigated stream (with 1, 3, 5, and 10 per cent probability of occurrence) are presented in such a manner that the ordinates of a flood wave nearest to the maximum discharge of given probability are multiplied by the ratio NQ(p%)Q_{max} (measured) – where NQ(p%) is the maximum discharge with probability of p% in the Gumbel system of coordinates, and Q_{max} (measured) is the peak value of the flood wave nearest to the NQ(p%) value. The four flood waves are cut off at the levels of 10 m^3/s, and then the areas under the respective curves represent the discharge volumes that should be stored. The curves of K (shown on the right hand side of Figure 9.3) can be constructed using a simple graphical procedure. This is a rather simplified method that can best be used in the case of small rivers and brooks.

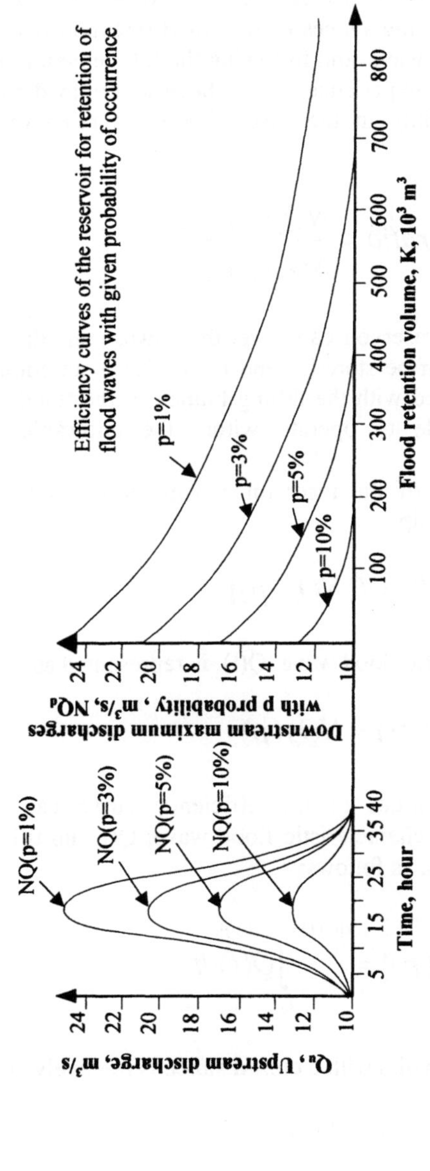

Figure 9.3. Construction of the efficiency curves for the Volgyseg flood retention reservoir – in the case of regulated sluice

9.2.2. THE CASE FOR A CLOSED SLUICE

The operation of the regulated sluice can be very costly. Consequently, within the valley of small watercourses, the sluices of the flood retention reservoirs are typically closed after passing the flood wave, and following the full depletion of the reservoir.

From our past practical experience, it has been determined that the ratio of the regulated sluice is higher only in the case of a flood retention that satisfies the following condition:

$$ratio = \frac{NQ_u(p)}{NQ_d(p)} \leq 2 \tag{9.4}$$

It is apparent that the closed reservoir can lower the downstream discharges if the rising branch of the flood wave can be stored – and there also is an additional capacity for retaining discharges associated with the falling branch of the flood wave. Under such circumstances, it is possible to operate within the allowable maximum of the downstream discharges.

The efficiency function of the reservoir having closed sluice is given by the following functional relationship:

$$K = f[NQ_d(p)] \tag{9.5}$$

The volume of the characteristic flood wave, Q(t), is represented as follows:

$$Q(t) = NQ_d(p) \tag{9.6}$$

Hence, the corresponding values of the efficiency curve can be calculated by integrating the volume of the characteristic flood wave, Q(t), up till the corresponding discharge of the falling branch, as follows:

$$K[NQ_d(p)] = \int_{t_o}^{t[Q(t)=NQ_d(p)]} Q(t)dt \tag{9.7}$$

Overall, the operation of reservoirs with a closed sluice is relatively simple.

9.2.3. THE CASE FOR AN OPENED SLUICE

Reservoirs with the most complex operation usually are those with an opened, narrow sluice – or a spillway with fixed sill. In such cases, the arriving flood wave will meet an opened sluice – or in the case of a complex operation, the flood wave will reach the water level of the spillway, causing a banking up and reducing the downstream discharges. The outcome is that, one part of the flood wave will be stored in the reservoir and the water level will increase. Consequently, there is an increase of the head at the sluice and the height of overflow at the spillway – with a concomitant growing of the outflow. The growing outflow is not counterbalanced by the regulation of the sluice – and this 'self-adjusting hydraulic process' should be analyzed in the dimensioning of the sluice, and then on that basis, one can determine the efficiency function of the flood retention reservoir. The hydraulic process can be analyzed by using the basic equations of hydraulics, the time series of inflow, the topographic equations, and the hydraulic relationships of the water-carrying capacity of the sluice and the spillway.

From the equation of continuity, the storage equation can be derived – as follows:

$$Q_u(t)dt - Q_d(t)dt = \Delta S(t) \tag{9.8}$$

where $Q_u(t)$ is the upstream discharge, $Q_d(t)$ is the downstream discharge, and $\Delta S(t)$ is the internal storage. Here, the $Q_u(t)$ discharges can be determined from the characteristic flood wave. The value of the downstream discharges, $Q_d(t)$, depends on the size of the sluice and spillway, and also on the head at the sluice and the height of overflow – so that:

$$Q_d(t) = f[h(t)] \tag{9.9}$$

This is illustrated below with the case of a dike drain, for which:

$$Q = c \cdot \sqrt{2gh} \cdot A \tag{9.10}$$

where c is the hydromodulus, g is the gravity, h is the head, and A is the cross-sectional area of the dike drain. For the spillway of length L, we obtain:

$$Q = c \cdot L \cdot \sqrt{2g} \cdot h^{\frac{3}{2}} \tag{9.11}$$

In the above equations (9.10) and (9.11), the head, h, is a function of the stored water volume, S(t).

On the basis of topographic curves, the stored volumes, S(H), belonging to the given depths, H, and heads, h, can be determined as follows:

$$S(H) = f(H) = f(H_o + h) \tag{9.12}$$

where H_o is the level of the sluice or the spillway. From this curve, the stored volumes can be determined on the basis of the following inverse relationship:

$$h = f^{-1}(S) \tag{9.13}$$

The curve characterized by the relationship $K = f[Q_d^{max}(p)]$, and constructed on the basis of a geodetic survey, will provide the relationship between the necessary volume of the reservoir, K, and the depth, H. It is apparent that the same curve will determine the relationship between the stored volume, S(H), and the depth, H, of the reservoir with capacity, K (Figure 9.4).

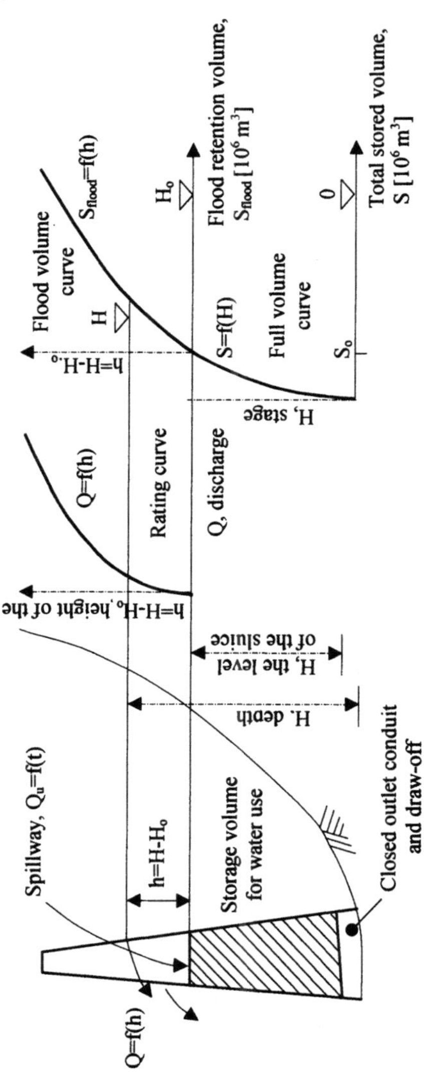

Figure 9.4. Flood retention of a multipurpose reservoir with fixed spillway

9.2.4. COMPARISON OF THE DIMENSIONING METHODS

The efficiency functions $K = f(NQ_d, p)$ of the three types of flood protection reservoirs discussed above are presented jointly in Figure 9.5 for a case of identical characteristic flood wave, $NQ_u(t)$. The efficiency functions showing the decrease of the flood wave will start and finish from the identical points given by the following respective equations:

$$NQ_d(p) = NQ_u(p); \quad K = 0 \tag{9.14}$$

$$NQ_d(p) = 0; \quad K = V_{fl}(p) \tag{9.15}$$

where V_{fl} is the full volume of flood wave (corresponding to the finishing point) with probability, p. It can be concluded that the reservoir volume necessary for the same retention of the flood wave is lower in the case of a closed sluice, in comparison to the case of an opened sluice. The difference between the efficiency of the different types of reservoirs will be especially significant in situations where the following is true:

$$n = \frac{NQ_u(p)}{NQ_d(p)} \leq 2 \tag{9.16}$$

The difference between the degrees of efficiency can be characterized by the ratios of volumes of the reservoirs that produce the same effect, as follows:

$$\frac{K_o[NQ_d(p)]}{K_r[NQ_d(p)]} \approx 2, \text{ if n} \leq 2 \tag{9.17}$$

and

$$\frac{K_c[NQ_d(p)]}{K_r[NQ_d(p)]} \approx 4, \text{ if n} \leq 2 \tag{9.18}$$

where K_o, K_c, K_r are the reservoir volumes in the case of opened, closed, and regulated sluices, respectively.

In conclusion, it is apparent that the downstream discharges can be significantly lowered by the use of regulated sluices. In the case of small brooks, the application of opened sluices may provide a more advantageous solution.

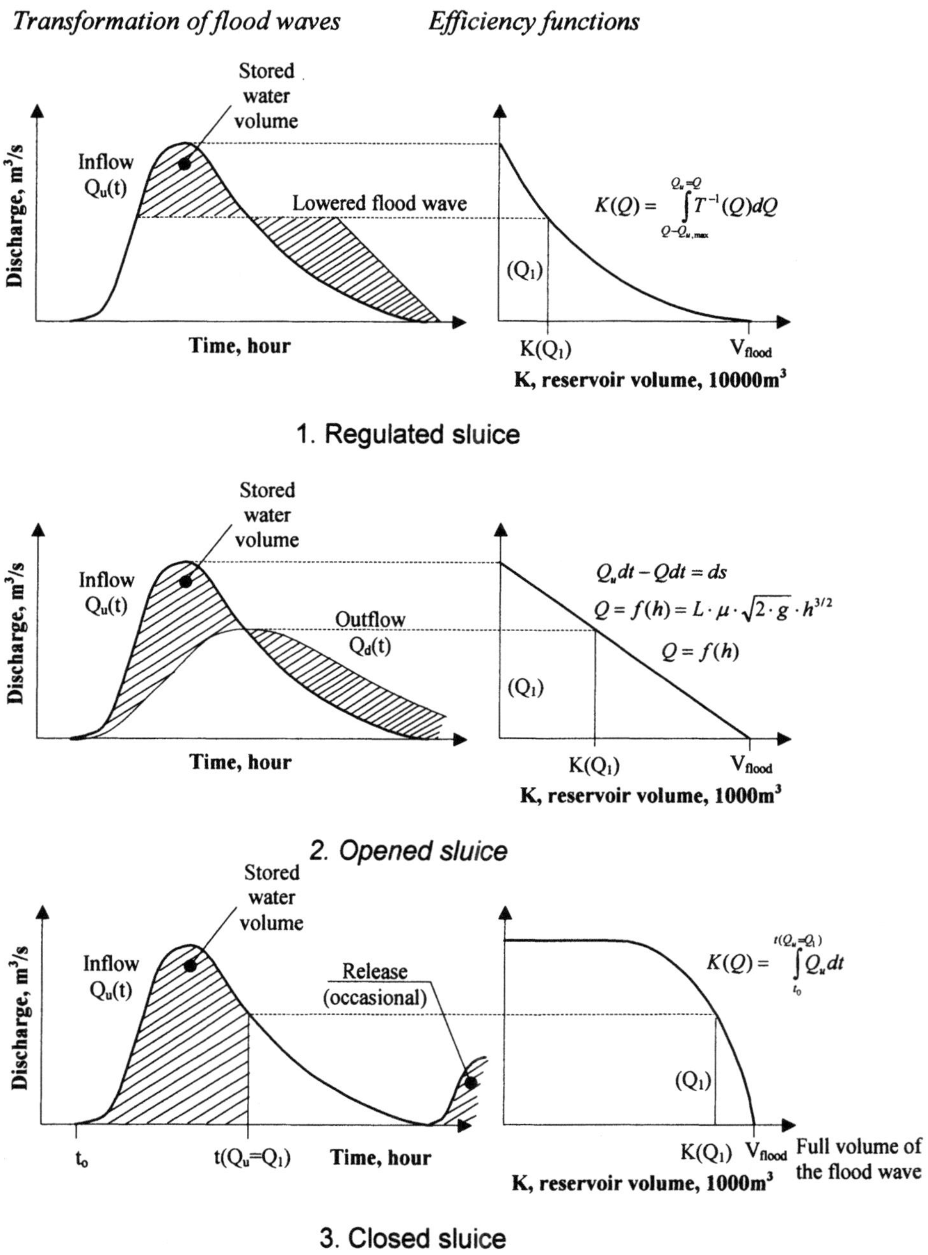

Figure 9.5. Transformation of flood waves and the efficiency functions

9.3. Approximate Evaluation of the Efficiency Function

The hydrological dimensioning process that fixes the main parameters of a planned reservoir usually is based on the results of a hydrologic survey of the watershed that has only limited measured data. As mentioned previously, the efficiency curve of the reservoir with opened sluice is between the curves of regulated and closed sluices. Consequently, it can be assumed that this curve (the shape of which depends on the character of the valley) can be approximated with a straight line – with K_l starting from the $[NQ_d(p) = NQ_u(p), K = 0]$ coordinates, and ending in the point $[NQ_d(p) = 0, K = V_{fl}(p)]$, where V_{fl} is the volume of the flood wave – as follows:

$$K_l[NQ_d(p)] = \frac{V_{fl}(p)}{NQ_u(p)} \cdot NQ_d(p) + V_{fl}(p) \tag{9.19}$$

or, equivalently:

$$K_l\left[NQ_d(p) = \frac{NQ_u(p)}{n}\right] = \frac{n-1}{n} \cdot V_{fl}(p) \tag{9.20}$$

where, in the case of an open sluice:

$$n = \frac{NQ_u(p)}{NQ_d(p)} \tag{9.21}$$

The above method (Bukovszky , 1989) can be used for the rapid evaluation of the storage possibilities. Assuming the following approximate relationship:

$$K = f(Q_d^{\max}) \cong S^{\max} \tag{9.22}$$

it can be concluded that, if there is a need for a storage volume, K_l, for the lowering of the flood wave in the case of an open sluice, where:

$$n = \frac{NQ_u(p)}{NQ_d(p)} \leq 2 \tag{9.23}$$

then the identical rate of lowering can be satisfied with a smaller regulated sluice represented by the following:

$$K_{r,l}\left[NQ_d(p) = \frac{NQ_u(p)}{n}\right] \cong \frac{n-1}{2 \cdot n} V_{fl}(p) \tag{9.24}$$

On the other hand, in the case of a closed sluice, the corresponding storage volume will be higher, given by:

$$K_{c,l}\left[NQ_d(p) = \frac{NQ_u(p)}{n}\right] \cong 2 \cdot K_l \cdot \frac{2\cdot(n-1)}{n} \cdot V_{fl}(p) \tag{9.25}$$

The above-indicated approximations can be applied only in situations where the condition defined by equation (9.23) is satisfied. If there is a need for a greater degree of flood retention, then the necessary storage will approach the volume of the reservoir with an open sluice. In the case of 'full retention', where $NQ_d(p) = 0$, and $n \to \infty$, the necessary reservoir capacity equals the full volume of the flood wave, viz.:

$$K_l = K_{r,l} = K_{c,l} = V_{fl}(p) \tag{9.26}$$

The above estimations are illustrated in Figure 9.6 for a case of triangular-shaped hydrographs. Here the ordinates produce a $V_{fl} = F = F_1+F_2+F_3+F_4 = 1$, and following relationships hold true:

$$(K = F_1):(F+1) = (1-Q)^2 : 1 \quad \text{..........for regulated sluice} \tag{9.27}$$

$$(F_1 + F_2):1 = (1-Q):1 \quad \text{...........for open sluice} \tag{9.28}$$

$$(F_1 + F_2 + F_3):1 = \left(1-\frac{2}{3}Q^2\right):1 \quad \text{..........for closed sluice} \tag{9.29}$$

It is noteworthy that, in designing large barrages or cofferdams, the hydrological dimensioning cannot be carried out by a rough estimation.

As an illustrative example of the process enumerated above, let us determine the main geometric characteristics of a flood retention reservoir for a small brook – with estimated values of $NQ_u(p = 3\ \%) = 18\ m^3/s$, and $NQ_u(p = 1\%) = 25.2\ m^3/s$. It is assumed in this problem that there are bridges with conveyance capacities of only 15 m^2/s – and our task is to determine the volume of the upstream reservoir that will be necessary for the appropriate flow regime of the existing bridges. The volume of the flood wave is given by (Bukovszky, 1989):

$$V_{fl}(p=1\%) = 1\,260\,000\ \text{m}^3 \tag{9.30}$$

and the retention rate is:

$$n = \frac{25}{15} \cong 1.67 \tag{9.31}$$

Thus, the required reservoir volume in the case of opened sluice is:

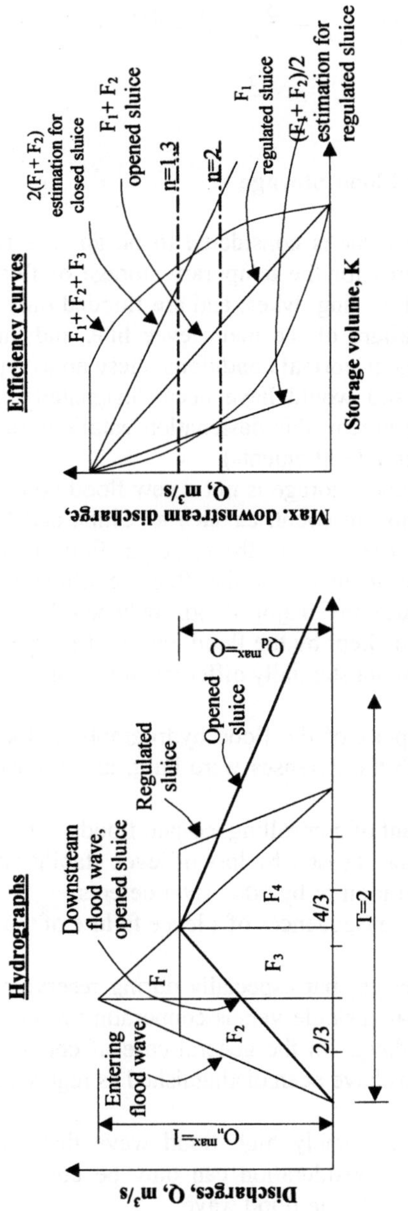

Figure 9.6. Constructed and estimated efficiency curves of the regulated, opened, and closed sluices of reservoirs

$$K_l = \frac{n-1}{n} \cdot V_{fl} = \frac{\frac{5}{3}-1}{\frac{5}{3}} \cdot 1\,260\,000 = 504\,000\ \mathrm{m}^3 \tag{9.32}$$

9.4. Emergency Flood Storage

An emergency reservoir is considered to be an area that has been made suitable by engineering measures for the temporary storage of flood flows. These reservoirs are usually surrounded by ring dykes, and are flooded only in extraordinary situations – to avert impending failure of the main levee line, and thus to control major losses and flood disasters. Under normal conditions, these areas typically serve agricultural and/or forestry purposes and would have been designated for such use by the responsible authorities of government; this designation entails limitations in land use (e.g., ban on building any residential settlements).

Indeed, emergency storage is not a new flood control method because it has a long history of application in countries such as China and India. In some countries, these reservoirs are classified among the principal flood control structures; for instance, in Hungary, they are included in the flood confinement plans of the district water authorities – because the major floods between 1925 and 1995 along several river sections could not be kept by the flood levees. Emergency flood storage may generally be warranted in four substantially different situations, namely:

(a) To lower the peak of the flood hydrograph at flood stages surpassing the design level for which the defenses were built, and for which they are capable of safely withstanding.
(b) As an instrument of controlling ice-jam floods – to avert impending levee failure.
(c) To prevent flood disaster by loss of levee stability owing to saturation from floods of extended duration or by other river defects.
(d) To control the consequences of a levee failure of the given river sections.

In reservoir design, but especially during reservoir operation, distinction must be made between isolated/single versus cooperating reservoirs (i.e., connected parallel or in series to each other). In the general case of cooperating reservoirs, three different hydrologic situations have been distinguished as regards emergency storage, namely:

1. The 'solitary' extremely high flood wave that has to be decreased, where the reservoir under consideration can only be counted upon if it is along the river section pertaining to the flood wave.
2. Flood emergency situation on several rivers, where flooding of all reservoirs within the entire river system may become necessary.
3. In the case when the emergency reservoir is at the confluence (delta) of two rivers, then it should be carefully designed to cater for both the 'solitary' flood wave and either of the two tributaries that conveys the larger flood volume above the particular flood level – whiles at the same time taking the additional water volume resulting from lowering the level on the other river into account.

Parallel or series connected reservoirs will also ensure an effective level control in these hydrological situations. In general, any decisions made about flooding the emergency reservoir is usually taken by considering the prevailing conditions. The decision maker has to take into account the potential economic as well as social consequences – recognizing that flooding the emergency reservoirs may entail considerable costs and losses.

9.4.1. IMPACTS OF EMERGENCY STORAGE IN THE RIVER SYSTEM

The potential impacts of emergency storage in the river system are illustrated in Figure 9.7. The magnitude and character of the impacts on the various sections of the river system will depend on the relative location of the section and the point of diversion.

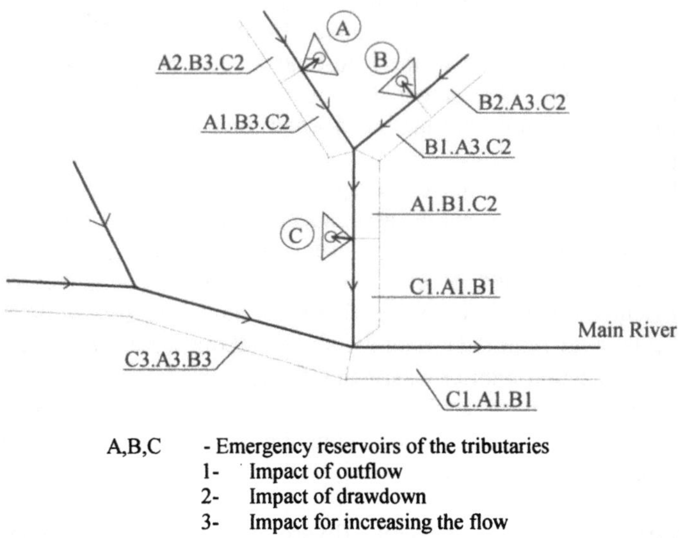

Figure 9.7. Impacts of emergency storage in the river system (after Szlavik, 1998)

Downstream of the diversion, the impact of emergency storage propagates along the particular branch of the river system and is felt over a long river section. The flow diverted into the reservoir results in lowering the hydrograph of the flood wave. Indeed, the discharge diverted into the reservoir may even be higher than the flood flow in the river – due to the local drawdown that may temporarily reverse the direction of flow over the downstream reach towards the diversion. Upstream of the diversion, the drawdown is a marked one that decreases rapidly with distance to become steady beyond a certain point.

Overall, the impact of diversion on the tributaries to the branch of the river system from which water is diverted is an indirect one – in that, as the water level in the main river is lowered, the slope and the velocity of flow in the tributaries is increased to convey the arriving flow at a lower water level. Thus, the impacts of each emergency reservoir on the various sections of the stream network must be carefully identified and quantified in advance. The results must be incorporated into the various operational scenarios; these scenarios should clearly state the situations in which the operation of a single emergency reservoir is ineffective, or will have only limited effects.

9.5. Design and Operation of an Emergency Flood Control Program

A general operation control model of the emergency flood reservoirs within the stream network, and the computations involved in operating the model, are shown in a block diagram (Figure 9.8). The main issues relevant to the design (dimensioning) and the operation of emergency flood reservoirs have been formulated by Szlavik (1998); the computations and evaluations involved can be classified into three key blocks – elaborated below.

- Block (A) contains the specialized hydrological preparatory operations necessary for decision-making. The time required for this sequence of operations may consume a considerable part of the lead-time for forecasting, especially on rivers with a flashy regime. The preparatory operations must therefore be started at a time when some of the information needed for decision-making on reservoir flooding is still pending.
- Block (B) contains the steps of selection, estimation, and computation of the emergency storage options and the assessment of the effects of flooding. Here, the analysis of past emergency storage events is necessary, and the preparatory operations must be planned, ordered, and executed in such a manner as to perform the most essential activities up to specific times – and the reservoir must be readied for flooding with the required reliability, albeit without superfluous expenditure.
- Block (C) is the action plan of emergency storage which must be reviewed, and if necessary, modified at regular intervals of time to take any advances in potential engineering measures (e.g., flood containment, drainage, control/abatement technologies, etc.) into account.

There currently are no general (hydrologic, hydraulic, economic, and environmental) methods for modeling the phenomena and processes involved in emergency flood storage. But it is essential to maintain continuous interaction between the hydrological forecasts and the analysis of emergency storage options.

Flooding an emergency reservoir must be preceded by a set of special technical preparations and a careful, thorough organization. The preparatory operations can be organized by using the critical path method of scheduling. In the case of the action plan, the duration of the critical path refers to the length of time during which the step-by-step sequence of preparatory operations must be started and completed prior to the computed point in time for flooding the emergency reservoir. Implementation of the action plan must be adjusted continuously in the light of the hydrological forecasts, and the preparations can be suspended, or abandoned at any time up to the commencement of an operation that has irreversible consequences (Szlavik, 1998).

Flooding an emergency reservoir is indeed the outcome of an extremely complex decision making process. Overall, a review of the relevant experiences gained during past emergency storage events in Central Europe (as an example) has indicated the need to set up a special system of hydrographic observations, measurements, and data communication. These should be laid down in guidelines – which go into force whenever the parameters of the particular flood wave indicate a potential need for emergency storage. Ultimately, concerted efforts of the local and national agencies controlling and implementing flood-fighting operations are essential in the final decision-making and execution process.

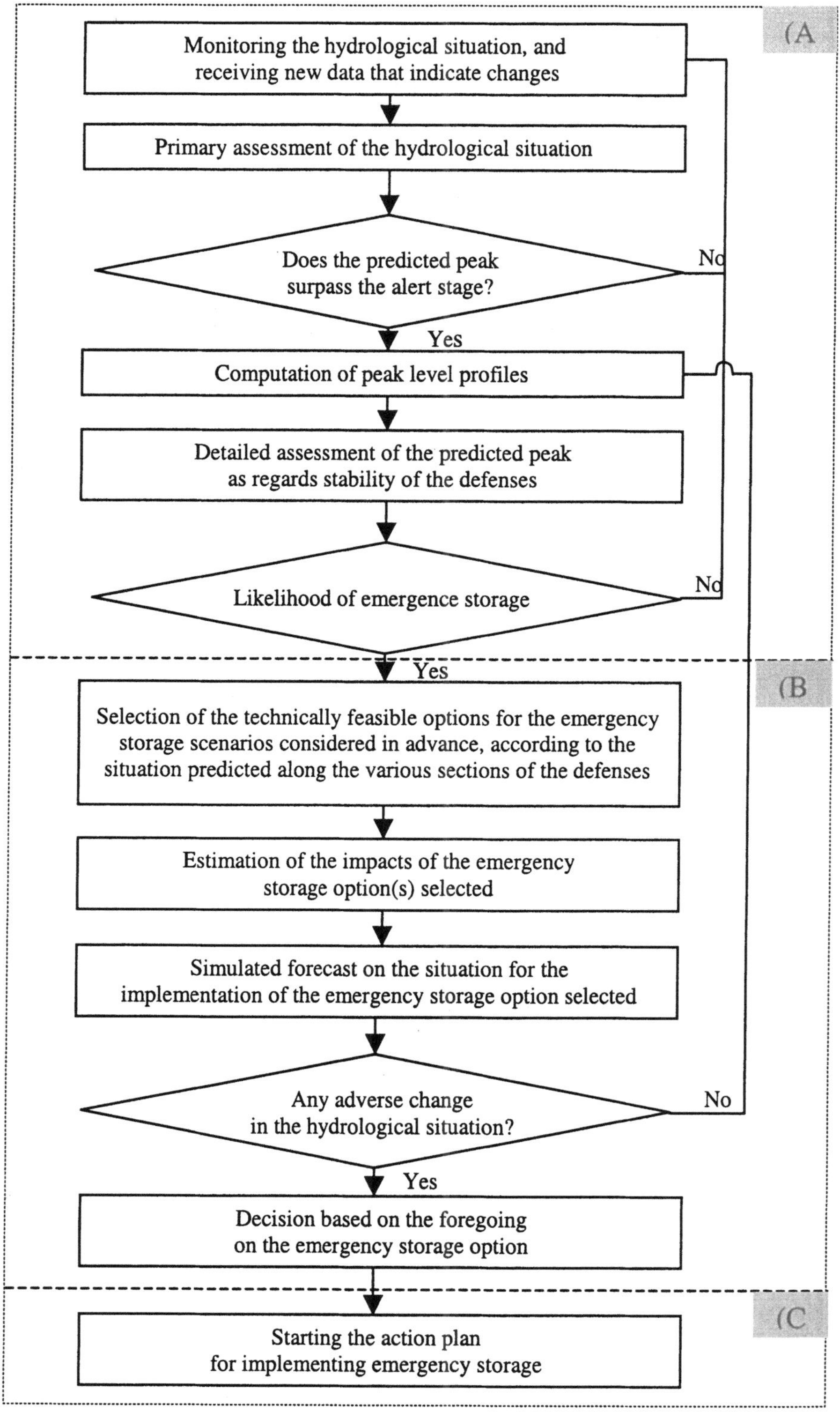

Figure 9.8. General operation control model of emergency reservoirs (after Szlavik, 1998)

Chapter 10

APPLICATION OF THE MORAN MODEL IN RESERVOIR STORAGE DESIGN

The application of the Moran theory (that is based on transition probabilities) to solve water storage-related problems seem to have found only limited applications in the past. A transition probability matrix shows the probability that a variable will assume any given value at the end of a time period, given that it has any specific value at the beginning of the period (McMahon and Mein, 1986). The reservoir states can be expressed in a transition matrix that shows the probability of finishing in a particular state as a function of the starting state. In view of the great potential utility of the transition probability matrix approach in the hydrological dimensioning of reservoirs, this chapter offers a discussion of some new efforts made in this direction – and we trust that water resources specialists will find these topics relating to the application of transition probabilities in the reservoir design effort to be of great interest.

10.1. The Moran Model

A theoretical outcome of the work by Moran (1959) is the recognition of the fact that the series of stored independent water volumes in time steps *Dt* can be characterized by a Markov-chain model. Moran proposed the discretization of the water volumes entering a reservoir of volume *K* that is used to meet a demand *M*. This means that the continuous random variables of the entering and stored volumes should be discretized for the multiple discrete values of *DV*, and for each element. Thence, the conditional probability distribution function, *F*, that relates the random variable, *S*, to the conditional probability distribution function of the stored water volumes in the reservoir (having volume, $K = k{\cdot}DV$ and satisfying demand, $M = m{\cdot}DV$), i.e.,

$$F(x/k,m) = P(S \le x \cdot DV/K = k \cdot DV, M = m \cdot DV) \qquad (10.1)$$

can be calculated from the probability distribution function of the entering water volumes, $Q_u(Dt)$, viz.:

$$H(y) = P\left[Q_u(Dt) \le y\right] \qquad (10.2)$$

The reservoir storage volume, K, and the water demand in each reservoir, M, are defined as follows:

$$K = k \cdot DV \text{ where, } k = 2, 3, ..., k_{max} \tag{10.3}$$

$$M = m \cdot DV \text{ where, } m = 1, 2, ..., (k-1) \tag{10.4}$$

It is obvious that the analysis of the lower limit (i.e., $k = 1$) is not relevant – because in the case of the smallest supply (i.e., $m = 1$), the probability of emptying of the reservoir equals unity (i.e., $p = 1$). On the other hand, the maximum value of m will at all times be lower than the capacity of the investigated reservoir, K.

The arriving water volume in the t-th year, $Q_u(Dt)$, is a random variable with distribution given by:

$$H(x) = P[Q_u(Dt) \leq x] \tag{10.5}$$

The continuous probability distribution function of the entering water volumes should be discretized and substituted by a step function (using the DV unit volumes) – as presented in the Figure 10.1, and detailed as follows:

$$\begin{aligned}
p_o &= P[Q_u(Dt) \approx 0] = P\left[Q_u(Dt) \leq \frac{1}{2}DV\right] = H(\frac{1}{2}DV) \\
p_1 &= P[Q_u(Dt) \approx 1DV] = P\left[\frac{1}{2}DV \leq Q_u(Dt) \leq (1+\frac{1}{2})DV\right] \\
&= H\left[(1+\frac{1}{2})DV\right] - H(\frac{1}{2}DV) \\
p_2 &= P[Q_u(Dt) \approx 2DV] = P\left[(1+\frac{1}{2})DV \leq Q_u(Dt) \leq (2+\frac{1}{2})DV\right] \\
&= H\left[(2+\frac{1}{2})DV\right] - H\left[(1+\frac{1}{2})DV\right] \\
p_3 &= P[Q_u(Dt) \approx 3DV] = P\left[(2+\frac{1}{2})DV \leq Q_u(Dt) \leq (3+\frac{1}{2})DV\right] \\
&= H\left[(3+\frac{1}{2})DV\right] - H\left[(2+\frac{1}{2})DV\right] \\
&\vdots \\
p_i &= P[Q_u(Dt) \approx iDV] = P\left[(i-1+\frac{1}{2})DV \leq Q_u(Dt) \leq (i+\frac{1}{2})DV\right] \\
&= H\left[(i+\frac{1}{2})DV\right] - H\left[(i-\frac{1}{2})DV\right]
\end{aligned} \tag{10.6}$$

Indeed, the stepped and discretized function of the probabilities *DV* can also be fitted to the curve of theoretical distribution of the measured data series.

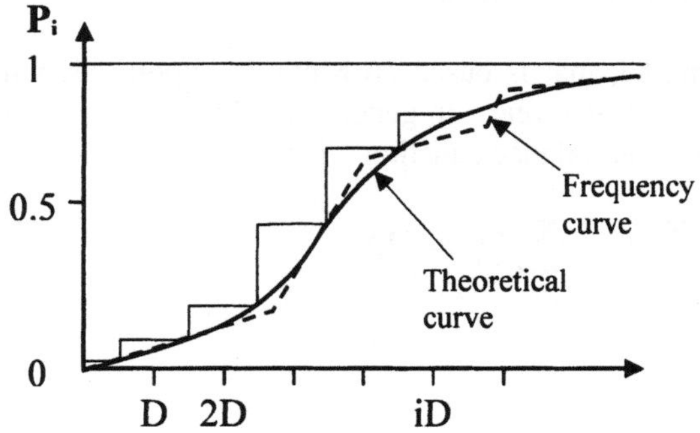

Figure 10.1. Discretization of the entering water volumes, and the fitting of the theoretical probability distribution

The numerical solution to the integral of equation (10.1) can be obtained by using the estimated probabilistic values of the discretized arriving unit water volumes, *DV*. To obtain the solution, several conditions have to be formulated; the most important condition – i.e., the independence of the arriving water volumes – can be determined by controlling the minimum length of time, *Dt*, such that:

$$P\{Q_u(jDt)/Q_u[(j-1)Dt]\} = P[Q_u(jDt)] \quad (10.7)$$

In a relatively long time period, there is a need to preliminarily fix the joint time conditions of the entered and consumed water volumes. The basic hypothesis of Moran was that, in the first wet period, Dt_1, of the entire *Dt* time interval, there is only inflow to the reservoir without water supply; and in the second dry period, Dt_2, there is no inflow, and only the collected water volumes are used for water supply. Here, the assumption is that $Dt = 1$ year, and the following relationships hold true:

$$Dt = Dt_1 + Dt_2 \quad (10.8)$$

$$Q_u(Dt_1) = Q_u(Dt), \text{ but } \mathrm{M}(\mathrm{Dt}_1) = 0 \quad (10.9)$$

$$Q_u(Dt_2) = 0, \text{ but } \mathrm{M}(\mathrm{Dt}_2) = \mathrm{M}(\mathrm{Dt}) \quad (10.10)$$

In general, the Moran hypothesis is justifiable only in the case of one-year time units – although it may also be applied to reservoir operation studies involving smaller time periods.

It is noteworthy that, another comparable 'extreme' view is the hypothesis of Lloyd (1974), in which the inflow and water supply is uniform within the *Dt* time period. In any case, the Moran model typically will result in an over-sizing (in the interest of improved reliability), whereas the Lloyd model tends to result in an under-dimensioning scenario.

10.2. A Proposed Mathematical Model

A mathematical model, that is based on Moran's hypothesis, will be introduced here using a time step of $Dt = 1$ year. In general, the basis of dimensioning of reservoirs is the analysis of the water balance equation, viz.:

$$\sum_{Dt} Q_u - \sum_{Dt} Q_d = \underset{Dt}{D}S \tag{10.11}$$

where,

$\sum_{Dt} Q_u = Q_u(Dt)$ is the sum of discharges arriving into the reservoir ($Dt = 1$ year),

$\sum_{Dt} Q_d = M(Dt)$ is the water demand in the Dt time period,

$\underset{Dt}{DS}$ is the change of stored water volume.

The water balance equation (10.11) of the reservoir is the basis of the Rippl's method, and also the dimensioning of the flood retention reservoirs. On the basis of Moran's hypothesis, the stored water volume in the reservoir can be calculated – and it is also possible to evaluate the theoretical probability distribution of the stored water volume.

Let us assume that in the year, *t*, preceding the dry period with supply, there is a storage volume, S(t-Dt), that remains in the reservoir; and in the next wet period of the year, *t*, the volume of entering water is $Q_u(Dt)$. Then, at the end of the wet period, the stored volume will be equal to the sum of the two water volumes – else an overflow will occur. Consequently, the water volumes in the reservoir are:

$$S^*(t) = S(t-Dt) + Q_u(Dt), \text{if } S(t-Dt) + Q_u(Dt) < K = kDV \tag{10.12}$$

or

$$S^*(t) = K = k \cdot DV, \text{if } S(t-Dt) + Q_u(Dt) \geq K = k \cdot DV \tag{10.13}$$

On the spillway, the overflow, T_f, is given by:

$$T_f(t) = 0, \text{if } S(t-Dt) + Q_u(Dt) < K + k \cdot DV \tag{10.14}$$

or

$$T_f(t) = [S(t-Dt) + Q_u(Dt)] - K, \text{if } S(t-Dt) + Q_u(Dt) \geq K = kDV \tag{10.15}$$

The above relationships can be compressed into single formulae for the storage volume and overflow, respectively. The stored water volume at the end of the wet period is given by:

$$S^*(t) = \min\{[S(t-Dt) + Q_u(Dt)], K\} \tag{10.16}$$

and the overflow in the year, *t*, is given by:

$$T_f(t) = \max\{([S(t-Dt) + Q_u(Dt)] - K), 0\} \tag{10.17}$$

The stored volume at the end of the dry period, S(t), is the difference between the water volume at the end of the wet period, S*(t), and the supply, M = *m·DV,* otherwise the reservoir will be depleted. Thus:

$$S(t) = S^*(t) - M, \text{if } S^*(t) - M > 0 \tag{10.18}$$

or

$$S(t) = 0, \text{if } S^*(t) - M \le 0 \tag{10.19}$$

In the case of the emptying of the reservoir, the full water demand cannot be satisfied. The water demand shortfall, *dt*, can be defined by the following relationships:

$$dt = 0, \text{if } S^*(t) - M > 0 \tag{10.20}$$

or

$$dt = S^*(t) - M, \text{if } S^*(t) - M \le 0 \tag{10.21}$$

where *dt* is a negative value. Once again, the above relationships can be compressed into single formulae for the storage volume and supply deficit, respectively. Thus, after consumption, the stored water in the t-th year is:

$$S(t) = \max\{[S^*(t) - M], 0\} \tag{10.22}$$

and the occasional water shortage is:

$$d(t) = \min\{[S^*(t) - M], 0\} \tag{10.23}$$

The above relationships can be further combined into single formulae; the stored water volume (after the supply) at the end of t-th year, S(t), the overflow, and the water shortage are , respectively, given by:

$$S(t) = \max[\min\{[S(t-Dt) + Q_u(Dt)], K\} - M, 0] \tag{10.24}$$

$$T_f(t) = \max\{([S(t-Dt) + Q_u(Dt)] - K), 0\} \tag{10.25}$$

$$d(t) = \min[(\min\{[S(t-Dt) + Q_u(Dt)], K\} - M), 0] \quad (10.26)$$

In the relationship proposed by Moran, the capacity of the planned reservoir, K = $k \cdot DV$, and the demand, M = $m \cdot DV$, are fixed values. However, for the construction of the characteristic capacity function of the reservoir, different K = $k \cdot DV$ and M = $m \cdot DV$ series may be utilized, based on any assumed or determined cyclical variations. The probability distribution of the water volume stored in the reservoir at the end of the year, S(t), is determined by using equation (10.6) – in which the discharge, $Q_u(t)$, and the stored water volume, S(t-Dt), are independent random variables. A similar relationship exists between the stored S(t) and S(t-Dt) water volumes defined by equation (10.24) – which means that they represent a single step Markov chain.

10.2.1. DETERMINATION OF THE TRANSITION PROBABILITIES

The transition probabilities, P_{ij}, characterizing the Markov chain can be calculated from equation (10.24), as follows:

$$P_{ij} = P[S(t) = i \cdot DV / S(t-Dt) = j \cdot DV] \quad (10.27)$$

Here, the transition probability is the conditional probability of a water volume of S(t) = $i \cdot DV$ remaining in the reservoir at the end of t-th year, if in the previous year there remained a water volume of S(t-Dt) = $j \cdot DV$. These P_{ij} transition probabilities may be arranged in a [(k-m+1) x (k-m+1)] size matrix. It follows from equation (10.24) that, if the inflow is higher than the capacity of the reservoir, then the next period of supply without inflow, Dt_2, will start with a stored water volume of:

$$S^*(t) = K = k \cdot DV \quad (10.28)$$

At the end of that period, the maximum stored water volume would be:

$$S(t) = S^*(t) - M = K - M = k \cdot DV - m \cdot DV = (k-m) \cdot DV \quad (10.29)$$

It may happen that the reservoir is emptied at the end of a given year, so the remaining water volume is S(t) = $i \cdot DV$, i = 0,1,2,...(k-m) – which means that there is a need for the evaluation of the transition probabilities of the different (k-m+1) values. For example, the probability of emptying of the reservoir in the t-th year, assuming the reservoir has been emptied in the previous year is given by:

$$P_{oo} = P[S(t) = 0 / S(t-Dt) = 0] \quad (10.30)$$

These transition probabilities can be calculated based on equation (10.24) – and this can be satisfied only if:

$$Q_u(t) \le M = m \cdot DV \quad (10.31)$$

This means that, the P_{00} value calculated from the known, discretized probabilistic values of the inflow is:

$$P_{oo} = P[S(t) = 0 / S(t - Dt) = 0] = P[Q_u \leq m \cdot DV] = \sum_{i=0}^{m} p_i \tag{10.32}$$

where p_i is the probability of the inflow, calculated on the basis of equation (10.6), so that:

$$p_i = P[Q_u(Dt) = i \cdot DV] = P\left[\left(i - \frac{1}{2}\right)DV < Q_u \leq \left(i + \frac{1}{2}\right)DV\right]$$

$$= H\left[\left(i + \frac{1}{2}\right)DV\right] - H\left[\left(i - \frac{1}{2}\right)DV\right] \tag{10.33}$$

where H(x) is the probability distribution of the inflow. The transition probability, P_{01}, is determined in a similar manner, so that:

$$P_{o1} = P[S(t) = 0 / S(t - Dt) = 1 \cdot DV] \tag{10.34}$$

which is the probability that the reservoir will be emptied in the t-th year if the remaining water volume in the previous year equals to $1 \cdot DV$ units. Again, these transition probability values can be calculated based on equation (10.24), as follows:

$$S(t) = 0 = \max[\min\{S(t - Dt) = 1 \cdot DV + Q_u(t), K\} - M, 0] \tag{10.35}$$

under the following conditions:

$$Q_u \leq M - DV = (m - 1) \cdot DV \tag{10.36}$$

Thence, the P_{01} transitional probability is calculated based on known, discretized values of the probabilities of the inflow, as follows:

$$P_{01} = P[Q_u \leq (m - 1) \cdot DV] = \sum_{i=0}^{m-1} p_i \tag{10.37}$$

The general P_{oj} element of the first row of the transitional probability matrix is determined by using the following relationship:

$$S(t) = 0 = \max[\min\{S(t - Dt) = j \cdot DV + Q_u(t), K\} - M, 0] \tag{10.38}$$

If in the previous year, $(t\text{-}Dt)$, the stored water volume was $j \cdot DV$, then the reservoir will be depleted only if the inflow of the t-th year was:

$$Q_u \geq M - j \cdot DV = (m - j) \cdot DV \tag{10.39}$$

Thence, the P_{oj} transition probability is calculated from the known, discretized probabilistic values of the inflow, as follows:

$$P_{oj} = P[Q_u \leq (m-j) \cdot DV] = \sum_{i=0}^{m-j} p_i \tag{10.40}$$

Now, the transition probability:

$$P_{10} = P[S(t) = 1 \cdot DV / S(t-Dt) = 0] \tag{10.41}$$

is the probability that at the end of the t-th year there will remain an S(t) = $1 \cdot DV$ unit of water volume in the reservoir if the reservoir has been emptied in the previous year. In terms of storages, the following relationship is true:

$$S(t) = 1 \cdot DV = \max[\min\{S(t-Dt) = 0 + Q_u(t), K\} - M, 0] \tag{10.42}$$

only if the following condition is met:

$$Q_u(t) = M + DV = (m+1) \cdot DV \tag{10.43}$$

Thus, the value of the transition probability, P_{10}, can be calculated based on known, discretized probabilistic values of inflow, as follows:

$$P_{10} = P[Q_u(t) = (m+1) \cdot DV] = p_{m+1} \tag{10.44}$$

where p_{m+1} is the calculated probability of the inflow.

The probability that at the end of the t-th year, there will remain an S(t) = $1 \cdot DV$ unit water volume in the reservoir if the remaining water volume at the end of the previous year was S(*t-Dt*) = $1 \cdot DV$ represents the transition probability that is defined by:

$$P_{11} = P[S(t) = 1 \cdot DV / S(t-Dt) = 1 \cdot DV] \tag{10.45}$$

Thus:

$$S(t) = 1 \cdot DV = \max[\min\{S(t-Dt) = 1 \cdot DV + Q_u(t), K\} - M, 0] \tag{10.46}$$

if:

$$Q_u(t) = M = m \cdot DV \tag{10.47}$$

Then the P_{11} transition probability is calculated from the known, discretized probabilistic values of the entering inflow, as follows:

$$P_{11} = P[Q_u(t) = m \cdot DV] = p_m \tag{10.48}$$

where p_m is the calculated probability of the inflow.

The probability that at the end of the t-th year there will remain an $S(t\text{-}Dt) = j{\cdot}DV$ water volume in the reservoir defines the transition probability, viz.:

$$P_{1j} = P[S(t) = 1 \cdot DV / S(t - Dt) = j \cdot DV] \tag{10.49}$$

Thus, the following equality relationship is realized:

$$S(t) = 1 \cdot DV = \max[\min\{S(t - Dt) = j \cdot DV + Q_u(t), K\} - M, 0] \tag{10.50}$$

only if:

$$Q_u(t) = M - j \cdot DV + DV = (m - j + 1) \cdot DV \tag{10.51}$$

In general, the transition probability, P_{1j}, can be calculated from the known discretized probabilistic values of the inflow, so that:

$$P_{1j} = P[Q_u(t) = m \cdot DV] = p_{m-j+1} \tag{10.52}$$

where p_{m-j+1} is the calculated probability of the inflow.

The probability that at the end of the t-th year there will remain a water volume of $S(t) = j{\cdot}DV$ in the reservoir, if it has been emptied in the previous year, is represented by the following transition probability:

$$P_{j0} = P[S(t) = j \cdot DV / S(t - Dt) = 0] \tag{10.53}$$

Thus, the following equality relationship:

$$S(t) = j \cdot DV = \max[\min\{S(t - Dt) = 0 + Q_u(t), K\} - M, 0] \tag{10.54}$$

is realized only if:

$$Q_u(t) = M + j \cdot DV = (m - j) \cdot DV \tag{10.55}$$

Consequently, the transition probability, P_{j0}, can be calculated from the known probabilistic values of the inflow, as follows:

$$P_{j0} = P[Q_u(t) = (m + j) \cdot DV] = p_{m+j} \tag{10.56}$$

Repeating the above-described procedure once again, the probability that at the end of the t-th year, the water volume of the reservoir will be equal to $S(t) = j{\cdot}DV$, if at the end of the previous year the remaining water volume was $S(t\text{-}Dt) = 1{\cdot}DV$, is given by the following transition probability:

$$P_{j1} = P[S(t) = j \cdot DV / S(t - Dt) = 1 \cdot DV] \tag{10.57}$$

This equality relationship:

$$S(t) = j \cdot DV = \max[\min\{S(t - Dt) = 1 \cdot DV + Q_u(t), K\} - M, 0] \tag{10.58}$$

is realized only if:

$$Q_u(t) = M + j \cdot DV - 1 \cdot DV = (m + j - 1) \cdot DV \tag{10.59}$$

Thus, the transition probability, P_{j1}, can be calculated from the known discretized probabilistic values of the inflow, as follows:

$$P_{j1} = P[Q_u(t) = (m + j - 1) \cdot DV] = p_{m+j-1} \tag{10.60}$$

Next, the probability that in the end of the t-th year the unit water volume of the reservoir is S(t) = $j \cdot DV$ if at the end of the previous year the remaining water volume equals S(*t-Dt*) = $i \cdot DV$ is represented by the following transition probability:

$$P_{j1} = P[S(t) = j \cdot DV / S(t - Dt) = i \cdot DV] \tag{10.61}$$

This equality relationship:

$$S(t) = j \cdot DV = \max[\min\{S(t - Dt) = i \cdot DV + Q_u(t), K\} - M, 0] \tag{10.62}$$

can be realized only if:

$$Q_u(t) = M + j \cdot DV - i \cdot DV = (m - j - i) \cdot DV \tag{10.63}$$

Thus:

$$P_{ji} = P[Q_u(t) = (m + j - 1) \cdot DV] = p_{m+j-i} \tag{10.64}$$

Similarly:

$$P_{jj} = P[S(t) = j \cdot DV / S(t - Dt) = j \cdot DV] \tag{10.65}$$

This equality relationship:

$$S(t) = j \cdot DV = \max[\min\{S(t - Dt) = j \cdot DV + Q_u(t), K\} - M, 0] \tag{10.66}$$

is valid only if:

$$Q_u(t) = m \cdot DV \tag{10.67}$$

which means that:

$$P_{jj} = p_m \tag{10.68}$$

It follows from the general relationship that if the sub-indexes of i and j were increased by the same number, g = 1,2,3,…, the result remains the same – so that:

$$P_{j+g,i+g} = P[S(t) = (j + g) \cdot DV / S(t - Dt) = (i + g) \cdot DV] \tag{10.69}$$

For the following equality relationship:

$$\begin{aligned} S(t) &= (j+g)\cdot DV \\ &= \max\left[\min\{S(t-Dt)=(i+g)\cdot DV + Q_u(Dt,K)\} - M, 0\right] \end{aligned} \quad (10.70)$$

the discretized probability of the inflow is given by:

$$P_{j+g,i+g} = P\left[Q_u = (m+j-i)\cdot DV\right] = p_{m+j-i} \quad (10.71)$$

where:

$$Q_u(Dt) = (m+j+g-i-g)\cdot DV = (m+j-i)\cdot DV \quad (10.72)$$

This means that the elements of the transitional probability matrix, P_{ij}, are identical in the case of each *i, j* = 1,2,……,(*k-m-1*) index pairs. That is, in the inner block of the transitional probability matrix (from the row having *i* = 1 index to the *i* = (*k-m-1*)), all 'inclined row' will contain the same elements – and this inner block is a Toeplitz-type matrix. The first element of the last row of the transitional probability matrix, given by:

$$P_{k-m,0} = P\left[S(t) = (k-m)\cdot DV / S(t-Dt) = 0\right] \quad (10.73)$$

is associated with the next situation, represented by:

$$S(t) = (k-m)\cdot DV = \max\left[\min\{S(t-Dt) = 0 + Q_u, K\} - M, 0\right] \quad (10.74)$$

This means that, in the wet period (before the following supply), the reservoir is filled – i.e., $S^*(t) = K = k\cdot DV$. This can occur with a previously depleted reservoir if:

$$Q_u(Dt) \geq K = k\cdot DV \quad (10.75)$$

with the following transition probability:

$$P_{k-m,0} = P\left(Q_u \geq k\cdot DV\right) = \sum_{i=k}^{\infty} p_i \quad (10.76)$$

Similarly, the general *j*-element of the last row of the matrix of the transition probability:

$$P_{k-m,j} = P\left[S(t) = (k-m)\cdot DV / S(t-Dt) = j\cdot DV\right] \quad (10.77)$$

is associated with the following situation:

$$S(t) = (k-m)\cdot DV = \max\left[\min\{S(t-Dt) = j\cdot DV + Q_u, K\} - M, 0\right] \quad (10.78)$$

This means that, in the wet period the reservoir has been filled, so that:

$$S^*(t) = K = k \cdot DV \tag{10.79}$$

which is possible if the following is true:

$$Q_u \geq K - j \cdot DV = (k-j) \cdot DV \tag{10.80}$$

and consequently, the following transition probability:

$$P_{k-m,j} = P\left(Q_u \geq (k-j) \cdot DV\right) = \sum_{i=k-j}^{\infty} p_i \tag{10.81}$$

The transitional probability matrix for $K > 2M$ are presented in Box 10.1. The result is a stochastic matrix that contains all the possible positions of the system, and where the sum of probabilities equals to unity. Knowing the transition probabilities, i.e.,

$$P_{i,j} = P\left[S(t) = i \cdot DV / S(t - Dt) = j \cdot DV\right] \tag{10.82}$$

in relation to the of basic theorem of the total probability, i.e.,

$$P_i = \sum_{j=0}^{k-m} P_{i,j} \cdot P_j \tag{10.83}$$

we can calculate the necessary values of non-conditional probabilities, as follows:

$$P_0 = \sum_{j=0}^{k-m} P_{0,j} \cdot P_j = P_{0,0} \cdot P_0 + P_{0,1} \cdot P_1 + P_{0,2} \cdot P_2 + \ldots + P_{0,k-m} \cdot P_{k-m}$$
$$P_1 = \sum_{j=0}^{k-m} P_{1,j} \cdot P_j = P_{1,0} \cdot P_0 + P_{1,1} \cdot P_1 + P_{1,2} \cdot P_2 + \ldots + P_{1,k-m} \cdot P_{k-m}$$
$$\vdots$$
$$P_i = \sum_{j=0}^{k-m} P_{i,j} \cdot P_j = P_{i,0} \cdot P_0 + P_{i,1} \cdot P_1 + P_{i,2} \cdot P_2 + \ldots + P_{i,k-m} \cdot P_{k-m}$$
$$P_{k-m} = \sum_{j=0}^{k-m} P_{k-m,j} \cdot P_j = P_{k-m,0} \cdot P_0 + P_{k-m,1} \cdot P_1 + P_{k-m,2} \cdot P_2 + \ldots + P_{k-m,k-m} \cdot P_{k-m} \tag{10.84}$$

Finally, on the basis of the theorem of total probability, we have $(k\text{-}m\text{+}1)$ linear equations for the determination of the non-conditional probabilities, $P_0, P_1, \ldots P_{k\text{-}m}$. Re-arrangement of the equations produces a homogenous system of equations, as follows:

Number of the stored DV water units at the end of the year i	Number of the stored DV water units at the end of the previous year														
	j=0	j=1	j=2	j=3	...	j=j	...	j=i	...	j=m-1	j=m	j=m+1	...	j=k-m-1	j=k-m
I=0	$P_{o,o}=\sum_{i=0}^{m} p_i$	$P_{o,1}=\sum_{i=0}^{m-1} p_i$	$P_{o,2}=\sum_{i=0}^{m-2} p_i$	$P_{o,3}=\sum_{i=0}^{m-3} p_i$	...	$P_{o,j}=\sum_{i=0}^{m-j} p_i$	...	$P_{o,i}=\sum_{i=0}^{m-i} p_i$	...	$P_{o,m-1}=\sum_{i=0}^{1} p_i$	$P_{o,m}=p_0$	$P_{o,m+1}=0$	...	$P_{o,k-m-1}=0$	$P_{o,k-m}=0$
I=1	$P_{1,0}=p_{m+1}$	$P_{1,1}=p_m$	$P_{1,2}=p_{m-1}$	$P_{1,3}=p_{m-2}$	...	$P_{1,j}=p_{m+1-j}$	...	$P_{1,i}=p_{m+1-i}$	...	$P_{1,m-1}=p_2$	$P_{1,m}=p_1$	$P_{1,m+1}=p_0$	...	$P_{1,k-m-1}=0$	$P_{1,k-m}=0$
I=2	$P_{2,0}=p_{m+2}$	$P_{2,1}=p_{m+1}$	$P_{2,2}=p_m$	$P_{2,3}=p_{m-1}$	...	$P_{2,j}=p_{m+1-j}$	...	$P_{2,i}=p_{m+2-i}$	...	$P_{2,m-1}=p_3$	$P_{2,m}=p_2$	$P_{2,m+1}=p_1$	...	$P_{2,k-m-1}=0$	$P_{2,k-m}=0$
I=3	$P_{3,0}=p_{m+3}$	$P_{3,1}=p_{m+2}$	$P_{3,2}=p_{m+1}$	$P_{3,3}=p_m$	...	$P_{3,j}=p_{m+1-j}$	...	$P_{3,i}=p_{m+3-i}$	...	$P_{3,m-1}=p_4$	$P_{3,m}=p_3$	$P_{3,m+1}=p_2$	...	$P_{3,k-m-1}=0$	$P_{3,k-m}=0$
...	...	...	...	...	...	...	...	...	...	...	...	...	...	...	...
...	...	...	...	...	...	...	...	...	...	...	...	...	...	...	...
I=j	$P_{j,0}=p_{m+i}$	$P_{j,1}=p_{m+i-1}$	$P_{j,2}=p_{m+i-2}$	$P_{j,3}=p_{m+i-3}$	...	$P_{j,i}=p_m$	...	$P_{j,i}=p_{m+j-i}$	...	$P_{j,m-1}=p_{j+1}$	$P_{j,m}=p_j$	$P_{j,m+1}=p_{j-1}$	...	$P_{j,k-m-1}=p_{2m+j-k+1}$	$P_{j,k-m}=p_{2m+j-k}$
...	...	...	...	...	...	...	...	...	...	...	...	...	...	...	...
...	...	...	...	...	...	...	...	...	...	...	...	...	...	...	...
I=i	$P_{i,0}=p_{m+i}$	$P_{i,1}=p_{m+i-1}$	$P_{i,2}=p_{m+i-2}$	$P_{i,3}=p_{m+i-3}$	...	$P_{i,j}=p_{m+i-j}$	...	$P_{i,i}=p_m$	...	$P_{i,m-1}=p_{i+1}$	$P_{i,m}=p_i$	$P_{i,m+1}=p_{i-1}$	...	$P_{i,k-m-1}=p_{2m+i-k+1}$	$P_{i,k-m}=p_{2m+i-k}$
...	...	...	...	...	...	...	...	...	...	...	...	...	...	...	...
...	...	...	...	...	...	...	...	...	...	...	...	...	...	...	...
i=m-1	$P_{m-1,0}=p_{2m-1}$	$P_{m-1,1}=p_{2m-2}$	$P_{m-1,2}=p_{2m-3}$	$P_{m-1,3}=p_{2m-4}$	...	$P_{m-1,j}=p_{2m-1-j}$	...	$P_{m-1,i}=p_{2m-1-i}$	...	$P_{m-1,m-1}=p_m$	$P_{m-1,m}=p_{m-1}$	$P_{m-1,m+1}=p_{m-2}$	...	$P_{m-1,k-m-1}=p_{3m-k}$	$P_{m-1,k-m}=p_{3m-k-1}$
I=m	$P_{m,0}=p_{2m}$	$P_{m,1}=p_{2m-1}$	$P_{m,2}=p_{2m-2}$	$P_{m,3}=p_{2m-3}$	...	$P_{m,j}=p_{2m-j}$	...	$P_{m,i}=p_{2m-i}$	...	$P_{m,m-1}=p_{m+1}$	$P_{m,m}=p_m$	$P_{m,m+1}=p_{m-1}$	...	$P_{m,k-m+1}=p_{3m-k+1}$	$P_{m,k-m}=p_{3m-k}$
i=m+1	$P_{m+1,0}=p_{2m+1}$	$P_{m+1,1}=p_{2m}$	$P_{m+1,2}=p_{2m-1}$	$P_{m+1,3}=p_{2m-2}$	...	$P_{m+1,j}=p_{2m+1-j}$	...	$P_{m+1,i}=p_{2m+1-i}$	...	$P_{m+1,m-1}=p_{m+2}$	$P_{m+1,m}=p_{m+1}$	$P_{m+1,m+1}=p_m$	...	$P_{m+1,k-m+1}=p_{3m-k+2}$	$P_{m+1,k-m}=p_{3m-k+1}$
...	...	...	...	...	...	...	...	...	...	...	...	...	...	...	...
...	...	...	...	...	...	...	...	...	...	...	...	...	...	...	...
i=k-m-1	$P_{k-m-1,0}=p_{k-1}$	$P_{k-m-1,1}=p_{k-2}$	$P_{k-m-1,2}=p_{k-3}$	$P_{k-m-1,3}=p_{k-4}$	...	$P_{k-m-1,j}=p_{k-1-j}$	...	$P_{k-m-1,i}=p_{k-1-i}$	...	$P_{k-m-1,m-1}=p_{k-m}$	$P_{k-m-1,m}=p_{k-m-1}$	$P_{k-m-1,m+1}=p_{k-m-2}$	...	$P_{k-m-1,k-m-1}=p_m$	$P_{k-m-1,k-m}=p_{m-1}$
i=k-m	$P_{k-m,o}=\sum_{i=k}^{\infty} p_i$	$P_{k-m,1}=\sum_{i=k-1}^{\infty} p_i$	$P_{k-m,2}=\sum_{i=k-2}^{\infty} p_i$	$P_{k-m,3}=\sum_{i=k-3}^{\infty} p_i$	....	$P_{k-m,j}=\sum_{i=k-j}^{\infty} p_i$	...	$P_{k-m,i}=\sum_{i=k-i}^{\infty} p_i$	...	$P_{k-m,m-1}=\sum_{i=k-m+1}^{\infty} p_i$	$P_{k-m,m}=\sum_{i=k-m}^{\infty} p_i$	$P_{k-m,m+1}=\sum_{i=k-m-1}^{\infty} p_i$		$P_{k-m,k-m-1}=\sum_{i=m+1}^{\infty} p_i$	$P_{k-m-1,k-m}=\sum_{i=m}^{\infty} p_i$

Box 10.1. The Moran theory probability matrix

$$
\begin{aligned}
&(P_{0,0}-1)\cdot P_0+P_{0,1}\cdot P_1+P_{0,2}\cdot P_2+...+P_{0,k-m}\cdot P_{k-m}=0\\
&P_{1,0}\cdot P_0+(P_{1,1}-1)\cdot P_1+P_{1,2}\cdot P_2+...+P_{1,k-m}\cdot P_{k-m}=0\\
&P_{2,0}\cdot P_0+P_{2,1}\cdot P_1+(P_{2,2}-1)\cdot P_2+...+P_{2,k-m}\cdot P_{k-m}=0\\
&\cdot\\
&\cdot\\
&P_{j,0}\cdot P_0+P_{j,1}\cdot P_1+P_{j,2}\cdot P_2+...+(P_{j,j}-1)\cdot P_j+...+P_{j,k-m}\cdot P_{k-m}=0\\
&\cdot\\
&\cdot\\
&P_{k-m,0}\cdot P_0+P_{k-m,1}\cdot P_1+P_{k-m,2}\cdot P_2+...+(P_{k-m,k-m}-1)\cdot P_{k-m}=0
\end{aligned}
\tag{10.85}
$$

This can be characterized by using the following product of matrices:

$$\overline{\overline{P_{i,j}}}\cdot\overline{P_i}=0 \tag{10.86}$$

where $\overline{\overline{P}}_{i,j}$ is the matrix of transitional probabilities of the conditions of the reservoir in the case of K > 2*M*, characterized by equation (10.81), and $\overline{P}_i$ is the vector of non-conditional probabilities or the eigenvector of the transitional probability matrix. The theorem of total probabilities, i.e.,

$$P_i=\sum_{j=0}^{k-m}P_{i,j}\cdot P_j \tag{10.87}$$

can also be characterized by the product of matrices, as follows:

$$\overline{\overline{P}}_{i,j}\cdot\overline{P}=\overline{P} \tag{10.88}$$

Also, the matrix of the homogenous linear system of equations having (*k-m+1*) x (*k-m+1*) order is the difference of the $\overline{\overline{P}}_{i,j}$ transitional probability matrix and the unit matrix (having the same order), i.e.,

$$\overline{\overline{P}}{}^{o}_{i,j}=\overline{\overline{P}}_{i,j}-\overline{\overline{E}} \tag{10.89}$$

This means that, there is a possibility of a non-trivial solution of the homogenous linear system of equations. Following the known procedures of matrix algebra, it can be shown that a non-homogeneous linear system of equations can be derived whose solution represent the probabilities, as follows:

$$p_i=P\left[Q_u(Dt)=i\cdot DV\right] \tag{10.90}$$

$P_{0,0}-1$	$P_{0,1}$	$P_{0,2}$	...	$P_{0,k-m-1}$	...	P_0	0
$P_{1,0}$	$P_{1,1}-1$	$P_{1,2}$	...	$P_{1,k-m-1}$	...	P_1	0
$P_{2,0}$	$P_{2,1}$	$P_{2,2}-1$	...	$P_{2,k-m-1}$	...	P_2	0
$P_{3,0}$	$P_{3,1}$	$P_{3,2}$	...	$P_{3,k-m-1}$	...	P_3	0
...	...	...	...	...	...	...	...
$P_{k-m-1,0}$	$P_{k-m-1,1}$	$P_{k-m-1,2}$	...	$P_{k-m,k-m-1}$	...	$P_{k-m-1}-1$	0
1	1	1	...	1	...	P_{k-m}	1

(10.91)

10.2.2. MODEL APPLICATION – AN EXAMPLE

A simplified Moran model can be applied to the preliminary evaluation of the relationship between water demand and the reservoir volume. This situation is demonstrated below by an example problem, utilizing discharge measurements from the 1974 – 1999 inflow data for the Csaszarviz reservoir in Hungary. Table 10.1 presents the discharges (Q, m^2/sec) and volumes (V, 10^6 m^3) of the dry period, wet period, and the average yearly values for this project – with the wet period lasting from the 1st of November to the 1st of May.

In the case of a homogenous Markov-chain model, the transitional probabilities of the reservoir falling into different conditions of fullness are characterized as follows:

$$\Pi = \begin{vmatrix} p_{00} & p_{01} & \cdots & p_{0,K-M} \\ p_{10} & p_{11} & \cdots & p_{1,K-M} \\ \cdot & \cdot & \cdots & \cdot \\ \cdot & \cdot & \cdots & \cdot \\ \cdot & \cdot & \cdots & \cdot \\ p_{K-M,0} & p_{K-M,1} & \cdots & p_{K-M,K-M} \end{vmatrix}$$

where K is the capacity (volume) of the reservoir and M is the supply. The numerical calculations are carried out in accordance with the steps listed below.

1. Let us consider a situation where the volume of the reservoir $K = 10 \cdot 10^6$ m^3 and the supply $M = 8 \cdot 10^6$ m^3. The size of the matrix, $(K-M)+1 = 3$, is given by:

$$\Pi_3 = \begin{vmatrix} p_{00} & p_{01} & p_{02} \\ p_{10} & p_{11} & p_{12} \\ p_{20} & p_{21} & p_{22} \end{vmatrix}$$

Table 10.1. Discharge measurements (from 1974-1999 inflow data) for the Csaszarviz reservoir in Hungary

Year	Dry period		Wet period		Yearly values	
	Qjm³/3	Vj,10⁶m³	Qi,m³/s	Vi,10⁶m³	Q,m³/s	V,10⁶m³
1974	0,047	0,641	0,850	13,403	0,897	14,144
75	0,079	1,246	0,747	11,779	0,826	13,025
76	0,196	3,091	0,963	15,185	1,159	18,276
77	0,174	2,744	1,445	22,785	1,619	25,529
78	0,150	2,365	1,598	25,157	1,748	27,563
79	0,164	2,586	0,609	19,603	0,773	22,189
80	0,392	6,181	1,420	22,390	1,812	28,517
1981	0,111	1,750	1,955	30,826	2,066	32,576
82	0,184	2,901	1,814	28,603	1,998	31,504
83	0,106	1,673	0,494	7,789	0,600	9,460
84	1,170	2,744	0,792	12,209	0,571	7,071
85	0,096	1,514	1,508	23,776	1,604	25,296
86	0,112	1,766	0,710	11,195	0,822	12,961
87	0,006	0,095	1,862	29,360	1,868	29,455
88	0,135	2,129	0,482	7,600	0,617	9,729
89	0,040	0,631	0,418	6,591	0,458	7,222
90	0,024	0,378	0,976	15,390	1,000	15,768
1991	0,240	3,784	1,102	17,376	1,342	21,160
92	0,136	2,144	0,677	10,075	0,813	12,819
93	0,169	2,665	1,330	20,971	1,499	23,636
94	0,179	2,822	0,443	6,985	0,622	9,807
95	0,224	3,532	0,874	13,781	1,098	17,313
96	0,184	2,901	1,585	24,992	1,769	27,893
97	0,151	2,381	1,065	16,793	1,216	19,174
98	0,169	2,665	0,723	11,400	0,892	14,065
99	0,073	1,151	0,662	10,438	0,735	11,589

(Here Q_j, Q_i, Q are the discharges of the dry, wet, and yearly periods; and V_j, V_i, V are the corresponding water volumes.)

2. In order to calculate the p_{00} value, take the *m* number of the years of wet periods when the inflow volumes $V_i \leq 8 \cdot 10^6$ m³. In this case, $m = 4$, and the relative frequency (probability) is:

$$p_{00} = \frac{m}{n} = \frac{4}{26} = 0.154$$

3. Calculate the value of p_{01} – by taking the number of years where the inflow volumes $V_i \leq 9 \cdot 10^6$ m³. Here $m = 0$, and relative frequency (probability) is:

$$p_{01} = \frac{m}{n} = \frac{0}{26} = 0$$

4. Calculate the value of P_{02} – by taking the number of years where $V_i \geq 10 \cdot 10^6$ m³. Here $m = 22$, and the relative frequency (probability) is:

$$p_{02} = \frac{m}{n} = \frac{22}{26} = 0.846$$

5. Continuing the above calculations produces the following results:

$$\Pi_3 = \begin{vmatrix} 0.15 & 0.00 & 0.85 \\ 0.08 & 0.08 & 0.84 \\ 0.00 & 0.08 & 0.92 \end{vmatrix}$$

6. Calculate the values of the absolute probabilities by raising the matrix to the dth power (where $\pi^d = \pi^{d-1}$). In our case, $d = 16$, and thence:

$$\Pi_3^{16} = \begin{vmatrix} p_0 = 0.008 & p_1 = 0.079 & p_2 = 0.913 \\ p_0 = 0.008 & p_1 = 0.079 & p_2 = 0.913 \\ p_0 = 0.008 & p_1 = 0.079 & p_2 = 0.913 \end{vmatrix}$$

where the p_0 (and p_1, p_2) are the in the absolute probabilities of falling into the fullness for 0, 1, or 2 of the reservoir, independently from the previous state. The characteristic value is the absolute probability, $p_0 = 0.008$, which is the probability of emptying of the reservoir. This also means that the length of the 'memory' of the investigated system is $d = 16$ years.

7. Continuing the above calculations results in the following matrix:

Volume of the reservoir [K, $10^6 m^3$]	Water supply [M, $10^6 m^3$]	Absolute probabilities of emptying, p_0 (%)
10	8	0.008
10	9	0.160
15	9	0.001
15	10	0.026
20	11	0.014
20	12	0.069
20	13	0.166
25	11	0.000
25	12	0.019
25	13	0.072

8. Using a linear interpolation between the p_0 values, a graphical K = f(M) relationship (Figure 10.2) is constructed for the different p_0 = (0.01 to 0.2) values. The reliability of supply is equal to $p = (1-p_0)$ %. The typical values used in Europe are:

- p = 99 % for drinking water supply,
- p = 95-98 % for industrial water supply,
- p = 70-85 % for irrigation in the semi-arid zone,
- p = 80-95 % for irrigation in the arid zone.

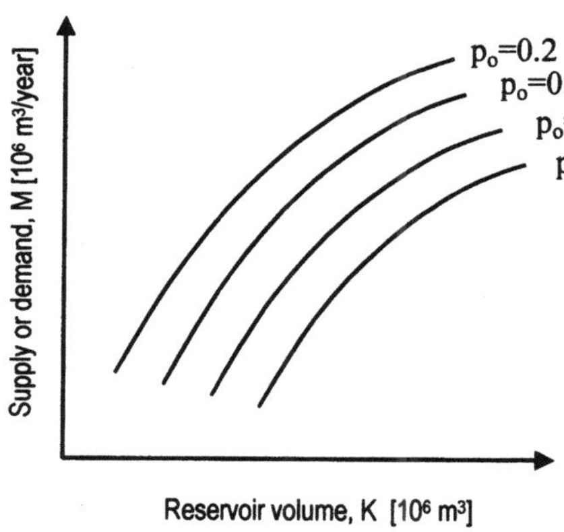

Figure 10.2. Schematic representation of the K = f(M) relationship

10.3. A Probabilistic Model for the Determination of the Reservoir Efficiency Function

Based on our practical experience from the hydrological design of storage reservoirs, we recognize great potentials in the utilization of the original Moran-type probabilistic model for determining the characteristic efficiency function that is represented by:

$$K = f\left(M = m \cdot DV, P\right) \tag{10.92}$$

The basic characteristics of the model formulation and the modeling process are discussed below.

10.3.1. THE BASIC HYPOTHESES

Discretizing the probability distribution function of the continuous, independent random variables:

$$F(x) = P\left[Q_u(t) \leq x\right] \tag{10.93}$$

by using *DV* unit volumes, the probability distribution of the water volumes entering into the reservoir is determined as follows:

$$p_i = P[Q_u(t) = i \cdot DV] = F\left[\left(i + \frac{1}{2}\right) \cdot DV\right] - F\left[\left(i - \frac{1}{2}\right) \cdot DV\right] \tag{10.94}$$

In terms of the Moran hypotheses, the discharges entering into the reservoir, $Q_u(t)$, in the first time period, Dt-Dt_1, are given by:

$$Q_u(Dt - Dt_1) \geq 0 \text{ and } Q_u(Dt_1) = 0 \tag{10.95}$$

and the water supply in the second time period, Dt_1, by the following:

$$M(Dt - Dt_1) = 0 \text{ and } M(Dt_1) = m \cdot DV > 0 \tag{10.96}$$

10.3.2. THE BASIC RELATIONSHIPS

1. The state of the reservoir before the designated supply is represented by:

$$S^*(t) = \min[S(t-1) + Q_u(t), K] \tag{10.97}$$

2. The state of the reservoir after the designated supply is represented by:

$$S(t) = \max\{[S^*(t) - M],0\} = \max\{(\min[S(t-1) + Q_u(t)], K) - M, 0\} \tag{10.98}$$

3. The shortfall in the designated supply is represented by:

$$d(t) = \min\{[S^*(t) - M],0\} = \min\{(\min[S(t-1) + Q_u(t)], K) - M, 0\} \tag{10.99}$$

4. The overflow from the filled reservoir is represented by:

$$T_f(t) = \max\{Q_u(t) - [K - S(t-1)],0\} \tag{10.100}$$

5. The probability distribution function of the stored water volumes at the end of time unit Dt, (i.e., after the supply) is given by:

$$P_i = P[S(t)/K = k \cdot DV, M = m \cdot DV] \tag{10.101}$$

where, $k = 2,3,\ldots,k_{max}$ and $m = 1,2,\ldots,(k-1)$. The task is to find the solution of the homogenous linear system of equations (given the stochastic vector of the conditional probabilities), in regards to the determination of the eigenvector of the transition probability matrix (having eigenvalue $\lambda = 1$), viz.:

$$\overline{\overline{P}}_{ij}(k,m) \cdot \overline{P}_i(k,m) = \overline{P}_i(k,m) \tag{10.102}$$

for which it can be shown that the columns of the matrix of absolute probabilities equals to the eigenvector (Prekopa, 1995), i.e.,

$$\lim_{r \to \infty} \overline{\overline{P}}_{ij}^{\,r} = \overline{P}_i^{\,j=0}, \overline{P}_i^{\,j=1}, \overline{P}_i^{\,j=2}, \ldots, \overline{P}_i^{\,j=k-m} \tag{10.103}$$

10.3.3. DETERMINATION OF THE TRANSITION PROBABILITY MATRIX

The $\overline{\overline{P}}_{ij}$ transition probability matrix of the stored water can be determined for each $K = k \cdot DV$, ($k = 2,3,\ldots,k^{max}$) and $M = m \cdot DV$, ($m = 1,2,\ldots,k-1$) by using the product of matrices, as follows:

$$\overline{\overline{Z}}(k,m) \cdot \overline{\overline{P}}_{i,j}(k \to \infty, m = 0) = \overline{\overline{P}}_{i,j}(k,m) \tag{10.104}$$

where

$\overline{\overline{Z}}(k,m)$ is a hyper-matrix in which the sizes of the:

- ➔ aggregated horizontal block vector, $\overline{\overline{1}}$, (containing two 1 elements),
- ➔ unit $\overline{\overline{E}}$ block matrix (containing $\underline{1}$ elements),
- ➔ horizontal block vector (containing $\underline{0}$ elements), and
- ➔ $\overline{\overline{0}}$ block matrix (containing two $\underline{0}$ elements),

are dependent only on the numbers of k and m, and are independent of the discharge values of the river. Thus,

$$\overline{\overline{Z}}(k,m) = \begin{array}{c} \\ 1 \\ 2 \\ 3 \\ \ldots \\ \ldots \\ k-m+1 \end{array} \begin{array}{c} \begin{array}{cccccccccccc} 1 & 2 & 3 & \ldots & m & m+1 & m+2 & \ldots & \ldots & \ldots & \ldots & \end{array} \\ \left[\begin{array}{cccccccccccc} 1 & 1 & 1 & \ldots & 1 & 0 & 0 & \ldots & 0 & 0 & 0 & \to\infty \\ 0 & 0 & 0 & \ldots & 0 & 1 & 0 & \ldots & 0 & 0 & 0 & \to\infty \\ 0 & 0 & 0 & \ldots & 0 & 0 & 1 & \ldots & 0 & 0 & 0 & \to\infty \\ \ldots & \ldots & \ldots & \ldots & \ldots & \ldots & \ldots & \ldots & \ldots & \ldots & \ldots & \\ 0 & 0 & 0 & \ldots & 0 & 0 & 0 & \ldots & 1 & 0 & 0 & \to\infty \\ 0 & 0 & 0 & \ldots & 0 & 0 & 0 & \ldots & 0 & 1 & 1 & \to\infty \end{array}\right] \end{array}$$

(10.105)

Consequently:

$$\overline{E}(m)\cdot\bar{0}(\infty),\overline{\overline{Z}}(k-m)=\overline{\overline{0}}(k-m-1\cdot m)\cdot\overline{\overline{E}}(k-m,k-m)\cdot\overline{\overline{0}}(k-m-2,\infty),\overline{E}(\infty) \tag{10.106}$$

The matrix of size (*k-m*,∞) (using 0,1 elements) is the function of the reservoir size (*k*) and the supply (*m*). The matrix in this case is:

$$\overline{\overline{P}}_{i,j}(k\to\infty,m=0)=\begin{bmatrix} p_0 & 0 & 0 & 0 & 0 & 0 & 0 & 0 & 0 & 0 & 0 & \to\infty \\ p_1 & p_0 & 0 & 0 & 0 & 0 & 0 & 0 & 0 & 0 & 0 & \to\infty \\ p_2 & p_1 & p_0 & 0 & 0 & 0 & 0 & 0 & 0 & 0 & 0 & \to\infty \\ p_3 & p_2 & p_1 & p_0 & 0 & 0 & 0 & 0 & 0 & 0 & 0 & \to\infty \\ p_4 & p_3 & p_2 & p_1 & p_0 & 0 & 0 & 0 & 0 & 0 & 0 & \to\infty \\ p_5 & p_4 & p_3 & p_2 & p_1 & p_0 & 0 & 0 & 0 & 0 & 0 & \to\infty \\ p_6 & p_5 & p_4 & p_3 & p_2 & p_1 & p_0 & 0 & 0 & 0 & 0 & \to\infty \\ p_7 & p_6 & p_5 & p_4 & p_3 & p_2 & p_1 & p_0 & 0 & 0 & 0 & \to\infty \\ p_8 & p_7 & p_6 & p_5 & p_4 & p_3 & p_2 & p_1 & p_0 & 0 & 0 & \to\infty \\ p_9 & p_8 & p_7 & p_6 & p_5 & p_4 & p_3 & p_2 & p_1 & p_0 & 0 & \to\infty \\ p_{10} & p_9 & p_8 & p_7 & p_6 & p_5 & p_4 & p_3 & p_2 & p_1 & p_0 & \to\infty \\ \downarrow & \downarrow & \downarrow & \downarrow & \downarrow & \downarrow & \downarrow & \downarrow & \downarrow & \downarrow & \downarrow & \\ \infty & \infty & \infty & \infty & \infty & \infty & \infty & \infty & \infty & \infty & \infty & \end{bmatrix} \tag{10.107}$$

which is a lower-grade triangular Toeplitz-type matrix – obtained from the probabilities of the discretized distribution of the natural inflow volumes (Rózsa, 1997).

10.3.4. DETERMINATION OF THE 'BEHAVIOR FUNCTION'

The 'behavior function' characterizing the probability distribution of the shortages in the reservoir, the stored water volumes, and the overflow from the filled reservoir can be determined by the hyper-matrix product that describes the behavior of reservoirs having different volumes, K = $m{\cdot}DV$ [$k = 2,3,\ldots,k_{max}$], and different water demands, M = $m{\cdot}DV$ [$m = 1,2,\ldots,(k\text{-}1)$]– described by the following expression:

$$\overline{\overline{[VIS]}}=\overline{\overline{[P]}}\cdot\overline{\overline{P}}(K\to\infty,M=0) \tag{10.108}$$

The stochastic row-vector, $\overline{\text{VIS}}$ (*k,m*), can be described as follows:

$$\overline{VIS}(k,m) = \overline{P}_i^{(r)}, \overline{P}_i, \overline{P}_i^{(tf)} \tag{10.109}$$

where

$$\overline{P}_i^{(r)} = P\left[M_r(t) = S^*(t) = i \cdot DV\right], \text{ if } i \le m \tag{10.110}$$

is the vector of the probabilities of the restricted $M_r \le M$ supply, and

$$\overline{P}_i = P\left[S(t) = i \cdot DV\right], \ i = 1,2,\ldots,k-m-1 \tag{10.111}$$

where S(t) is the remaining water volume after the supply, and

$$\overline{P}_i^{(tf)} = P\left[Tf = i \cdot DV\right], \ i = 0,1,2,3,\ldots \tag{10.112}$$

where Tf is the vector of probabilities of the overflow. It is apparent that

$$\sum_{i=0}^{m} P_i^{(r)} + \sum_{i=1}^{k-m-1} P_i + \sum_{i=0}^{\infty} P_i^{(tf)} = 1 \tag{10.113}$$

It is noteworthy that, it is possible to apply this methodology to reservoir dimensioning problems for periods of shorter duration. For instance, the transition probability matrices in the case of Dt less than one year (namely, seasons or groups of months) can be approximately determined using the Harris-type of non-stationary model (Prekopa, 1995) – represented by the following relationship:

$$\overline{\overline{Z}}(k, m(T)) \cdot \overline{\overline{P}}(T)(k \to \infty, m = 0) = \overline{\overline{P}}_{i,j}(T) \tag{10.114}$$

where the probability distribution of the natural discharges and the $M = m \cdot DV$ supply are the functions of the given time units, T. Thence, the corresponding behavior vector, VIS(T), can be obtained by using the method applied to the case of a steady state model, where $Dt = 1$ year.

10.4. Concluding Remarks

The objective of the structural analysis of different matrices that can be used in the dimensioning of the reservoirs is to help improve the overall design accuracy. The proposed methods elaborated above have been successfully applied in the dimensioning of several reservoirs in central Europe and elsewhere. Even so, it is understandable that this is not the last word in the processes involved in the development of theories in the dimensioning of reservoirs. For example, the possible application of the generalized random walk model, and the analysis of ergodicity of Markov-chains and Markov processes with finite or countably infinite states can be used in the determination of the probability of emptying and overspill of the reservoir, and/or for the calculation of the time of flooding (i.e., the duration of a flood-wave) (Reimann, 1989; V. Nagy, 1982). Ultimately, it is our hope that the ideas presented here in this book can be used for the development of further concepts, frameworks, and methodologies that will help to further improve the overall reservoir design process.

List of References and Bibliography

LITERATURE CITED

ACKERMANN, W., G. WHITE and E.B. WORTHINGTON (eds.) (1973) *Man-Made Lakes: Their Problems and Environmental Effects,* Geophysical Monograph 17, AGU, Washington, DC, USA.

ALBERTSON, M.M., L.S. TUCKER and D.C. TAYLOR (eds.) (1971) *Treatise on Urban Water Systems,* Colorado State University Press, Fort Collins, CO, USA.

ALEXANDER, G.N. (1962) The use of the Gamma Distribution in Estimating Regulated Output from Storages, *Civil Engineering Transaction – The Australian Institution of Engineers,* CE 4(1): 29-34.

ANDERSON, M.G. and T.P. BURT (ed.) (1985) *Hydrological Forecasting,* John Wiley and Sons, Ltd., England, UK.

ANDREANOV, V.G. (1975) *Meteorological and Hydrological Data Required in Planning the Development of Water Resources,* Operational Hydrology Report No.5 (WMO-No.419), Geneva, Switzerland.

ANG, A.H.-S. and W.H. TANG (1975) *Probability Concepts in Engineering Planning and Design, Vol. 1. - Basic Principles,* John Wiley and Sons, Inc., New York, USA.

ARIS, R. (1994) *Mathematical Modeling Techniques,* Dover Publications, Inc., New York, NY, USA.

ASKEW, A.J., F. GRECO and J. KINDLER (ed.) (1981) *Logistic and Benefits of Using Mathematical Models of Hydrologic and Water Resource Systems,* Selected Papers from an International Symposium, IIASA Proceedings Series 13, Pergamon Press.

ASKEW, A.J., W.W.-G. YEH and W.A. HALL (1971) Use of Monte Carlo Techniques in the Design and Operation of a Multi-purpose Reservoir System, *Water Resources Research,* 7(4), 819-826.

AUSTIN, R.J. (1998) Reaching the End of the Road. *Journal of International Water Power and Dam Construction,* 2:118-130.

BAGLEY, P.B. (1991) The Pulse Advantage and the Restoration of River-floodplain Systems. *Regulated Rivers: Research and Management,* 6:75-86.

BAKER, D., J. YOUNG and J.M. AROCENA (2000) An Integrated Approach to Reservoir Management: The Williston Reservoir Case Study, *Journal of Environmental Management,* 25(5): 565-578.

BÁRDOSSY, A., I. BOGARDI AND L. DUCKSTEIN (1987) Accuracy of Sediment Deposition Estimation. *Water Resources Research,* 23 (6):1021-1026.

BEARD, L.R. (1972) Hydrologic Data Management, in *Hydrologic Engineering Methods for Water Resources Development, Vol.2,* U.S. Army Corps of Engineers, Davis, California, USA.

BECKER, L. and W.W.-G. YEH (1974) Optimization of Real Time Operations of a Multi-Reservoir System *Water Resources Research,* 10(6), 1107-1112

BELLMAN, R. (1957) *Dynamic Programming,* Princeton University Press, Princeton, New Jersey, USA.

BENGTSSON, L. (1980) Moderate Storm Runoff from a Small Bog. *Vannet i Norden,* 1:27-41.

BERGA, L., J. YAGUE, J. CAJETE, F. GIRON AND J. MENDILUCE (1999) *Benefits and Concerns about Dams in Spain.* ICOLD, Antalya, pp. 79-95.

BERNIER, J. (1985) Prise en compte des incertitude d' information et des objectifs dans l'estimatic des modeles hydrologigues. *Revue Internationale des Sciences de l' Eau, Paris,* 1(4): 3-10.

BISWAS, A.K. (ed.) (1976) *Systems Approach to Water Management,* McGraw-Hill Book Co., New York.

BOGARDI, I. and F. SZIDAROVSZKY (1976) Application of Game Theory in Water Management, *Applied Mathematical Modelling,* Vol.1.

BOX, G.E.P. and G.M. JENKINS (1976) *Time Series Analysis: Forecasting and Control,* Holden-Day, Inc., San Francisco, CA, USA.

BREITENECKER, F., A. SCHMIDT and M. PESCHEL (1984) Simulation and Optimization of a Multipurpose Hydro-Energetic System Using Standard Simulation Software, *International Workshop on Applied Optimization Techniques in Energy Problems,* Linz, Austria.

BROTCHIE, J.F., R. SHARPE and A.R. TOACKLEY (1973) Econometric Model to Aid in Urban Planning Decisions, *Journal of Urban Planning and Development Division, ASCE,* 99, Paper No. 9587.

BRUCE, J.P. and R.H. CLARK (1966) *Introduction to Hydrometeorology,* Pergamon, Toronto, Canada.

BUKOVSZKY, G. (1989) Sizing of Small Flood Retention Reservoirs. *Hydraulic Engeneering,* 2: 10-15 Budapest, Hungary.

BURAS, N. (1972) *Scientific Allocation of Water Resources,* American Elsevier Publishing Co., Inc., New York, USA.

CARPENTER, P. (1970) *Cost-effectiveness as an Aid to Making Decisions in Education,* The Rand Corp., Santa Monica, Calif., USA.

CHIU, C.L. (1972) *Stochastic Methods in Hydraulics and Hydrology of Streamflow,* Geophysical Surveys, 1:61-84.

CHIU, C.L. (1978) *Application of Kalman Filter to Hydrology, Hydraulics and Water Resources,* Proceedings of AGU Chapman Conference, University of Pittsburgh, Pennsylvania, USA.

CHOW, V.T. (ed.) (1964) *Handbook of Applied Hydrology,* McGraw-Hill Book Co., New York, USA

CHUN HONG (1995). Controlling Reservoir Sedimentation in China, *Hydropower and Dams,* March, 1995, pp. 50-52.

CIRIANI, T.A., U. MAIONE and J.R. WALLIS (eds.) (1977) *Mathematical Models for Surface Water Hydrology,* Proceedings of Workshop at IBM Science Center, Pisa, Italy, John Wiley and Sons, Chichester, UK.

CLARKE, R.T. (1973) *Mathematical Models in Hydrology,* Irrigation and Drainage Paper 19, FAO/UN, Rome, Italy.

COHON, J.L. (1978) *Multiobjective Programming and Planning,* Mathematics in Science and Engineering, Vol. 140, Academic Press, Inc. New York, USA.

DAVID, L. (1975) *Water Resources Regulation,* Lecture Notes in Applied Hydrology, International Post-graduate Course on Hydrological Methods for Developing Water Resources Management, Research Institute for Water Resources Development, Budapest, Hungary.

DAVIDAN, I.N., L.I. LOPATUHIN and V.A. ROZSKOV (1989) *Vetrovoje Volnyenie (in Russian), (Wind induced waves),* Gidrometeoizdat, Leningrad, Russia.

DAVIS, C.V. and K.E. SORENSEN (1969) *Handbook of Applied Hydraulics, 3rd edition,* McGraw-Hill, New York, USA.

DETWYLER, T.R. (ed.) (1971) *Man's Impact on Environment,* McGraw-Hill, New York.

DOOGE, J.C.I. (1977) Problems and Methods of Rainfall-Runoff Modelling, in *Mathematical Models for Surface Water Hydrology,* edited by Ciriani *et. al.*, John Wiley and Sons, England, UK.

DOORENBOS, J. and W.O. PRUITT (1977) *Crop Water Requirements,* FAO Irrigation and Drainage Paper No.25, FAO/UN, Rome, Italy.

DYCK, S. and M. SCHRAMM (1967) *Application of Monte Carlo Methods to Reservoir Design,* Proceedings of International Hydrology Symposium, Volume 1, Fort Collins, Colo., USA.

EASTMANN, J. and C. ReVELLE (1973) Linear Decision Rule in Reservoir Management and Design, *Water Resources Research,* 9(1).

ECAFE (1962) *Proceedings of the Regional Symposium on Dams and Reservoirs (Tokyo),* Economic Commission for Asia and the Far East (ECAFE), Flood Control Series No.21., UN, New York.

EL-SHAARAWI, A.H. and S.R. ESTERBY (eds.) (1982) *Time Series Methods in Hydrosciences, Developments in Water Science 17,* Elsevier, Amsterdam, Holland.

FAIR, G.M., J.C. GEYER, and D.A. OKUN (1966) *Water and Wastewater Engineering, Volume 1: Water Supply and Wastewater Removal,* John Wiley and Sons, Inc., New York, USA.

FIERING, M.B. (1961) Queuing Theory and Simulation in Reservoir Design, *Journal of Hydraulic Division, ASCE,* 87(HY6), 39-69.

FIERING, M.B. (1967) *Streamflow Synthesis,* Harvard University Press, Cambridge, Mass., USA.

FIERING, M.B., J.J. HARRINGTON and R.J. De LUCIA (1967) *Water Resources Systems Analysis,* Resource Paper No.3, Policy Research and Coordination Branch, Dept. of Energy, Mines and Resources, Ottawa, Canada.

FIERING, M.B. and B.B. JACKSON (1971) *Synthetic Streamflows, Water Resources Monograph 1,* American Geophysical Union (AGU), Washington, DC.

FLEMING, G. (1974) *Deterministic Simulation in Hydrology,* American Elsevier, New York, USA.

FORD, L.R., Jr. and D.R. FULKERSON (1962) *Flow in Networks*, Princeton University Press, New Jersey, USA.

GODA, L. (1995) *Statistical and Probability Analysis of High Waters in Rivers and Creeks.* Statistical and Bayesian Methods in Hydrological Sciences. Conference, Paris, UNESCO, Sept. 1995. Proc. of the Conference, pp. 155-170.

GOODCHILD, M.F., L.T. STEYAERT, B.O. PARKS, *et al.* (eds.) (1996) *GIS and Environmental Modeling: Progress and Research Issues*, GIS World Books, Fort Collins, Colorado.

GOTTFRIED, B.S. and J. WEISMAN (1974) *Introduction to Optimization Theory*, Prentice-Hall International Series in Industrial and Systems Engineering.

GOTTSCHALK, L., G. LINDH and L. de MARE (eds.) (1971) *Stochastic Processes in Water Resources Engineering*, Water Resources Publication, Fort Collins, Colorado, USA.

GOUEVSKY, I.V. (1974) *On Optimum Control of Multireservoir Systems*, Research Report RR-74-25, Schloss-Laxenberg, Austria.

GOULD, B.W. (1961) Statistical Methods for Estimating the Design Capacity of Dams, *Journal of Institution of Engineers, Australia*, 33(12): 405-416.

GOULD, B.W. (1964) *Statistical Methods for Reservoir Yield Estimation*, Water Research Foundation of Australia, Report No.8.

HAAN, C.T. (1977) *Statistical Methods in Hydrology*, University of Iowa Press, Iowa, USA.

HAIMES, Y.Y. (1977) *Hierarchical Analyses of Water Resources Systems*, McGraw-Hill Book Co., New York, USA.

HAIMES, Y.Y., W.A. HALL and H.T. FREEDMAN (1975) *Multiobjective Optimization in Water Resources Systems*, Developments in Water Science 3, Elsevier, Amsterdam, Holland.

HALL, W.A. (1964) Optimum Design of a Multiple-Purpose Reservoir, *Journal of Hydraulic Division, ASCE*, HY4.

HALL, W.A. and J.A. DRACUP (1975) *Water Resources Systems Engineering*, McGraw-Hill Book Co., New York, USA.

HALL, W.A., A.J. ASKEW and W.W.-G YEH (1969) Use of Critical Period in Reservoir Analysis, *Water Resources Research*, 5(6), 1205-1215.

HALL, W.A. and D.T. HOWELL (1963) The optimization of single purpose reservoir design with the application of dynamic programming to synthetic hydrology samples, *Journal of Hydrology*, 1, 355-365.

HAMMERSLEY, J.M. and D.C. HANDSCOMB (1965) *Monte Carlo Methods*, Methuen and Co. Ltd., London, UK.

HARMS, A.G. and T.H. CAMPBELL (1967) An Extension to the Thomas-Fiering Model for the Sequential Generation of Streamflow, *Water Resources Research*, 3: 653.

HEC (1974) *HEC-3, Reservoir System Analysis for Conservation, Users Manual*, Hydrologic Engineering Center, U.S. Army Corps of Engineers, Davis, California.

HEC (1975) *Hydrologic Engineering Methods for Water Resources Development, Volume 8: Reservoir Yield*, Hydrologic Engineering Center, U.S. Army Corps of Engineers, Davis, California.

HEC (1976) *Hydrologic Engineering Methods for Water Resources Development, Volume 7: Flood Control by Reservoirs*, Hydrologic Engineering Center, U.S. Army Corps of Engineers, Davis, California.

HEC (1977) *Hydrologic Engineering Methods for Water Resources Management Volume 9: Reservoir System Analysis for Conservation*, Hydrologic Engineering Center, U.S. Army Corps of Engineers, Davis, California.

HOSHI, K., S.J. BURGES and I. YAMAOKA (1978) *Reservoir Design Capacities for Various Seasonal Operational Hydrology Models*, Proceedings of the Japanese Society of Civil Engineers, No.273, pp 121-134.

HOUCK, M.H. (1979) A Chance Constrained Optimization Model for Reservoir Design and Operation, *Water Resources Research*, 15(5), 1011-1016.

HOUCK, M.H. and J.L. COHON (1978) Sequential Explicitly Stochastic Linear Programming Models: A Proposed Method for Design and Management of Multiple Reservoir Systems, *Water Resources Research*, 14(2), 161-169.

HOUCK, M.H., J.L. COHON and C.S. ReVELLE (1980) Linear Decision Rule in Reservoir Management, 6, Incorporation of Economic Efficiency Benefits and Hydroelectric Energy Generation, *Water Resources Research*, 16(l) 196-200.

HOWE, C.W. (1971) *Benefit-Cost Analysis for Water System Planning*, AGU Water Resources Monograph No.2, Washington, D.C., USA.

HUFSCHMIDT, M.M. and M.B. FIERING (1966) *Simulation Techniques for Design of Water-Resource Systems*, Harvard University Press, Cambridge, Mass., USA.

HURST, H.E. (1951) Long term storage capacity of reservoirs, *ASCE Transactions,* 116: 770-799.

IAHS-UNESCO-WM0 (1974) *Mathematical Models in Hydrology,* Volume 12, Proceedings of the Warsaw Symposium, TAHS/UNESCO.

INSTITUTION OF CIVIL ENGINEERS (ICE) (1978) *Floods and Reservoir Safety: An Engineering Guide,* The Telford, Ltd., London, UK.

IVANOV, A.N. (1989) *Gidrologija, Regulirovanyije sztoka (in Russian), (Hydrology and regulation of the stream flow),* Moscow, Kolosz, Russia.

JACOBY, H.D. and D.P. LOUCKS (1972) The Combined Use of Optimization and Simulation Models in River Basin Planning *Water Resources Research,* 8(6) 1401-1414.

JAMES, L.D. and R.R. LEE (1971) *Economics of Water Resources Planning,* McGraw-Hill Book Co. New York, USA.

JANSEN, R.B. (ed.) (1988) *Advanced Dam Engineering for Design, Construction, and Rehabilitation,* Van Nostrand Reinhold, New York, NY.

JENKINS, G.M. and D.G. WATTS (1968) *Spectral Analysis and its Applications,* Holden-Day Inc., San Francisco, Calif., USA.

JENKINSON, A.F. (1969) *Estimation of Maximum Floods,* WMO Technical Note No.98, Geneva, Switzerland.

JENSEN, M.E. (ed.) (1999) *Evapotranspiration and Irrigation Water Requirements.* Manual No. 70, American Society of Civil Engineers (ASCE), New York.

JOHNSON, S., J. STEDINGER and K. TASCHURS (1991) Heuristic Operating Policies for Reservoir System Simulation, *Water Resources Research,* 27(5): 673-685.

KENDALL, M.G. and A. STUART (1968) *The Advanced Theory of Statistics, Volume 3,* Charles Griffin and Co. Ltd., London, UK.

KINDLER, J. (1977) The Monte Carlo Approach to Optimization of the Operation Rules for a System of Storage Reservoirs, *Hydrological Sciences Bulletin,* XXII, 1:203-214.

KINDLER, J. and C.S. RUSSELL (eds.) (1984) *Modeling Water Demands,* IIASA, Academic Press Inc., London, UK.

KINDLER, J. and H. ZUBEREK (1967) *On Some Multi-Site Multi-Season Streamflow Generation Models,* IIASA, Research Memorandum RM-76-76, Laxenburg, Austria.

KITE, G.W. (1977) *Frequency and Risk Analysis in Hydrology,* Water Resources Publications, Fort Collins, Colo., USA.

KLEMES, V. (1969) Reliability Estimates for a Storage Reservoir with Seasonal Input, *Journal of Hydrology,* 7, 198-216.

KLEMES, V. (1973) *Applications of Hydrology to Water Resources Management,* Operational Hydrology Report No.4, WMO No.356, Geneva, Switzerland.

KLEMES, V. (1974) The Hurst Phenomenon – A Puzzle? *Water Resources Research,* 10(4), 675-688.

KLEMES, V. (1977) Discrete Representation of Storage for Stochastic Reservoir Optimization *Water Resources Research,* 13(l), 149-158.

KLEMES, V. (1978) The Unreliability of Reliability Estimates of Storage Reservoir Performance Based on Short Streamflow Records, in *Risk and Reliability in Water Resources Volume 1,* University of Waterloo, Canada; pp. 127-136.

KLEMES, V. (1993) The Stochastic Reservoir: a comment. *Journal of Hydrology,* 165: 371-373.

KNOP, H. (ed.) (1974) *The Tennessee Valley Authority Experience,* Proceedings of the 1st Conference on Case Studies of Large-Scale Planning Projects, Volumes 1 & 2, IIASA, Laxenburg, Austria.

KOTTEGODA, K.T. (1970) *Statistical Methods of River Flow Synthesis for Water Resources Assessment,* Institution of Civil Engineers, Supplement 18.

KUIPER, E. (1971) *Water Resources Project Economics,* Butterworths, London, UK.

LAGLER, K.F. (ed.) (1969) *Man-Made Lakes: Planning and Development,* UNDP, FAO/UN, Rome, Italy.

LANE, W. (1979) *Applied Stochastic Techniques (Users Manual),* Bureau of Reclamation, Engineering and Research Center, Denver, Colorado, USA.

LANGBEIN, W.B. (1958) Queueing Theory and Water Storage, Proceedings, *Journal of Hydraulic Division, ASCE,* 84/HY5.

LANGBEIN, W.B. (1959) Queueing Theory and Water Storage, Proceedings, *Journal of Hydraulic Division, ASCE,* 85/HY8.

LASDON, L.S. (1970) *Optimization Theory for Large Systems,* MacMillan, New York, USA.

LINSLEY, R.K. and J.B. FRANZINI (1979) *Water-Resources Engineering, 3rd ed.,* McGraw-Hill Book Co., New York, USA.

LINSLEY, Jr. R.K., M.A. KOHLER and J.L. PAULHUS (1982) *Hydrology for Engineers, 3rd edition*, McGraw-Hill Book Co., USA.

LLOYD, E.H. (1967) Stochastic Reservoir Theory, *Advances in Hydrosciences*, 4, 281-339.

LLOYD, E.H. (1974) "What is and what is not a Markov Chain?" *Journal of Hydrology*, 22: 1-28.

LOUCKS, D.P. (1969) *Stochastic methods for analyzing river basin systems*, Technical Report 16, Water Resources Marine Science Center, Cornell University New York, USA.

LOUCKS, D.P., J.R. STEDINGER and D.A. HAITH (1981) *Water Resource Systems Planning and Analysis*, Prentice-Hall, Englewood Cliffs, New Jersey, USA.

MAASS, A., M.M. HUFSCHMIDT, R. DORFMAN, H.A. THOMAS, Jr., S.A. MARGLIN and G.M. FAIR (1962) *Design of Water-Resource Systems*, Harvard University Press, Cambridge, Mass., USA.

MAIDMENT, D.R. and V.T. CHOW (1981) Stochastic State Variable Dynamic Programming for Reservoir Systems Analysis, *Water Resources Research*, 17(6), 1578-1584.

MAIDMENT, D.R. *et al.* (1993) *Handbook of Hydrology*, McGraw-Hill, Inc. New York.

MAJOR, D.C. and R.L. LENTON (1979) Applied Water Resource Systems Planning, Prentice-Hall, Englewood Cliffs, New Jersey, USA.

MANDELBROT, B.B. and J.R. WALLIS. (1968) Noah, Joseph and Operational Hydrology. *Water Resources Research*, 4(5), 909-918.

MANDELBROT, B.B. and J.R. WALLIS (1969) Some Long-run Properties of Geophysical Records, *Water Resources Research*, 5(2), 321-340.

MARINO, M.A. and B. MOHAMMADI (1983) Reservoir Management: A Reliability Programming Approach, *Water Resources Research*, 19(3), 613-620.

MATALAS, N.C. (1967). Mathematical Assessment of Synthetic Hydrology, *Water Resources Research*, 3(4), 937-945.

MATALAS, N.C. and B. JACOBS (1964). *A Correlation Procedure for Augmenting Hydrologic Data*, U.S. Geological Survey Professional Paper 434-E.

MATALAS, N.C. and J.R. WALLIS (1971) *Correlation Constraints for Generating Processes*, Symposium on Mathematical Models in Hydrology, IAHS/UNESCO, Warsaw, Poland.

MAYS, L.W. (ed.) (1996) *Water Resources Handbook*, McGraw-Hill, New York, NY.

McMAHON T.A. and R.G. MEIN (1978) *Reservoir Capacity and Yield*, Developments in Water Science, 9, Elsevier, Amsterdam, Holland.

McMAHON, T.A. and R.G. MEIN (1986) *River and Reservoir Yield*, Water Resources Publications, Littleton, Colorado.

META SYSTEMS, INC. (1975) *Systems Analysis in Water Resources Planning*, Water Information Center Publishing, New York, USA.

MOBASHERI, F. (1970) A Two-stage Optimization Model for Design of a Multipurpose Reservoir *Water Resources Research*, 6(l), 22-31.

MOHAMMADI, B. and M.A. MARINO (1984) Reservoir Operation: Choice of Objective Functions, *Journal Water Resources Planning and Management, ASCE*, pp.15-29.

MORAN, P.A.P. (1959) *The Theory of Storage*, Methuen, London, UK.

MORIN, T.L. (1973) Pathology of a Dynamic Programming Sequencing Algorithm, *Water Resources Research*, 9(5), 1178-1185.

MOSAEDI, A. (1998) Hydological Sizing of Sedimenting Reservoir System for Irrigation and Water Supply, Ph.D. Dissertation, Technical University of Budapest, Dept. of Water Resources Engineering, pp.1-150.

MOSONYI, E., E. HAUCK and D. KOBERG (1980) Zur Verwendung des vermutlich gröszten Hochwassers (PMF) und der Beta-Verteilung des T-jahrlichen Hochwasserabflusses. Deutsche Gewasserkundliche Mitteilungen, 24. Ig. Heft 4/5.

MOSS, M.E., D.P. LETTENMAIER and E.F. WOOD (1978) On the Design of Hydrologic Data Networks, *EOS Transactions, AGU*, 59(8), 772-5.

MOUSAVI, H. and A.S. RAMAMURTHY (2000) Optimal Design of Multi-reservoir Systems for Water Supply, *Advances in Water Resources*, 23(2000): 613-624.

NAYAK, S.C. and S.R. ARORA (1971) Optimal Capacities for a Multiple Reservoir System Using the Linear Decision Rule, *Water Resources Research*, 7(3), 485-498.

NAYAK, S. and S.R. ARORA (1973) Capacity Decisions in a Multipurpose Multireservoir System, *Water Resources Research*, 9(5), 1166-1177.

NERC (1975) *Flood Studies Report*, National Environmental Research Council (NERC), London, UK.

NEUFVILLE, R. and D.H. MARKS (eds.) (1974) *Systems Planning and Design*, Prentice-Hall, Inc., USA.

O'CONNEL, P.E. and J.A. COLE (1980) *Reservoir Yield V: Reliability of Pumped Storage Yield*, Technical Report TR 164, Water Research Center (WRC), Medmenham-Stevenage-Swindom, England, UK.

PHATARFORD, R.M. (1976) Some Aspects of Stochastic Reservoir Theory, *Journal of Hydrology*, 30 199-217.

PLESKOV, J.F. (1985) *Regulirovanyije Recsnovo Sztoka (in Russian) (Regulation of Stream flow)*, Gidrometeoizdat, Leningrad, Russia.

PRATT, J.W., H. RAIFFA and R. SCHLAIFER (1965) *Introduction to Statistical Decision Theory*, McGraw-Hill, New York, USA.

PREKOPA, A. (ed.) (1973) *Inventory Control and Water Storage*, North-Holland Publishing Co., Amsterdam, Holland.

PREKOPA, A. (1995) *Stochastic Programming*. Kluwer Scientific Publishers, The Netherlands.

QUADE, E.S. (ed.) (1967) *Analysis for Military Decisions*, Rand-McNally, USA.

RAKOCZI, L. (1994) Damming of Low-Land Rivers, ICCORES Workshop, St. Petersburg, Proc. Chapt. 4, pp.1-25.

REIMANN, J. (1989) *Mathematical Statistics with Application in Flood Hydrology*, Acad. Publ., Budapest, Hungary.

ReVELLE, C. (1999) *Optimizing Reservoir Resources*, John Wiley & Sons, New York, NY, USA.

ReVELLE, C. and W. KIRBY (1970) Linear Decision Rule in Reservoir Management and Design 2, Performance Optimization, *Water Resources Research*, 6(4), 1033-1044.

ReVELLE, C., E. JOERES and W. KIRBY (1969) Linear Decision Rule in Reservoir Management and Design, 1, Development of the Stochastic Model, *Water Resources Research*, 5(4), 767-777.

RINALDI, S. (ed.) (1982) *Environmental Systems Analysis and Management*, North-Holland Publishing Co., Amsterdam, Holland.

RIPPL, W. (1883) *The Capacity of Storage Reservoirs for Water Supply*, Proceedings of the Institution of Civil Engineers, Vol. 71.

ROZSA, P. (1997) *Ordinal Measures of Association*, Publication of the Math. Institute, Hungarian Academy of Sciences, Vol. 4, pp.29-47.

SALAS, J.D., J.W. DELLEUR, V. YEVJEVICH and W.L. LANE (1980) *Applied Modeling of Hydrologic Time Series*, Water Resources Publication, Littleton, Colorado, USA.

SHWEIG Z. and J.A. COLE (1968) Optimal Control of Linked Reservoirs, *Water Resources Research*, 4(3) p. 479-497.

SIMONOVIC, S.P. and M.A. MARINO (1980) Reliability Programming in Reservoir Management, 1, Single Multipurpose Reservoir *Water Resources Research*, 16(5), 844-848

SIMONOVIC, S.P. and M.A. MARINO (1981) Reliability Programming in Reservoir Management, 2, Risk-loss functions, *Water Resources Research*, 17(4), 822-826.

SIMONOVIC, S.P. and M.A. MARINO (1982) Reliability Programming in Reservoir Management, 3, System of Multipurpose Reservoirs, *Water Resources Research*, 18(4), 735-743.

SIMONS, D.B. and F. SENTÜRK (1992) *Sediment Transport Technology and Water and Sediment Dynamics*, Water Resources Publication, Littleton, Colorado, USA.

SNIEDOVICH, M. (1979) Reliability-constrained Reservoir Control Problems, 1, Methodological Issues, *Water Resources Research*, 15(6): 1574-1582.

SOKOLOV, A.A. and T.G. CHAPMAN (1974) *Methods for Water Balance Computations*, Studies and Reports in Hydrology, 17, UNESCO, Paris, France.

STEDINGER, J.R. (1981) Estimating Correlations in Multivariate Streamflow Models, *Water Resources Research*, 17.

STEDINGER, J.R., B.F. SULE and D. PEI (1983) Multiple Reservoir System Screening Models, *Water Resources Research*, 19(6), 1383-1393.

STEPHENSON, D. (1970) Optimum Design of Complex Water Resource Projects *Journal of Hydraulic Division, ASCE*, HY6.

SVANIDZE, G.G. (1980) *Mathematical Modeling of Hydrologic Series, Translated from Russian*, Water Resources Publications, Fort Collins, Colo., USA.

SZAVARENSKIJ, A.D. (1951) *Regulirovanyije Recsnovo Sztoka vodohranilisesami (in Russian) (Regulation of the stream flow by the storage reservoirs)*, Izd-vo AH, CCCP, pp.1-236.

SZLAVIK, L (1998) Emergency Flood Storage, *Hydraulic Engineering*, No 1, pp.1-21, Budapest, Hungary.

TANG, W.H. and B.C. YEN (1972) Hydrologic and Hydraulic Design Under Uncertainties, in *Proceedings of the Internatioanl Symposium on Uncertainties in Hydrologic and Water Resources Systems, Volume 2*, pp. 868-882, Tucson, Arizona, USA.

THOMAS, D.M. and M.A. BENSON (1970) *Generalization of Streamflow Characteristics from Drainage Basin Characteristics,* USGS Water Supply Paper No. 1975.

THOMAS, H.A., Jr., and R.P. BURDERN (1963) *Operations Research in Water Quality Management,* Harvard Water Research Group, Cambridge, Mass., USA.

TROTT, W.J. and W.W.-G. YEH (1971) *Multi-level Optimization of a Reservoir System,* ASCE Annual and National Environmental Engineering Meeting, St. Louis, Missouri, 17 pp.

TROTT, W.J. and W.W.-G. YEH (1973) Optimization of Multiple Reservoir Systems, *Journal of Hydraulic Division, ASCE,* 99(HY10), 1865-1883.

UNESCO (1974) *Survey on Water Balance of Lakes and Reservoir of the World,* UNESCO, Paris, France.

UNESCO (1981) *Methods of Computation of the Water Balance of Large Lakes and Reservoirs, Volume 1: Methodology,* Studies and Reports in Hydrology, 31. Paris, France.

UNESCO (1982) *Methods of Hydrological Computations for Water Projects,* Studies and Reports No. 38, UNESCO, Paris, France.

UNESCO-WMO-IAHS (1974) *Design of Water Resources Projects with Inadequate Data, Volume 2,* Proceedings of the Madrid Symposium, June 1973, Contribution to the IHD Studies and Reports in Hydrology, 16.

USACERC (1977) *Shore Protection Manual.* U.S. Army Coastal Engineering Research Center (USACERC), Fort Belvoir, Viginia, USA.

USBR (1977) *Design of Small Dams,* U.S. Bureau of Reclamation, Dept. of the Interior, Washington, D.C.

VALENCIA, R. and J.C. SCHAAKE, Jr. (1973) Disaggregation Processes in Stochastic Hydrology, *Water Resources Research,* 9(3), 580-585

VANONI V. (ed.) (1976) *Sedimentation Engineering,* ASCE, New York, USA.

VIESSMAN, W., Jr., J.W. KNAPP, G.L. LEWIS and T.E. HARBAUGH (1977) *Introduction to Hydrology, 2nd edition,* Dun-Donnelley Publishing Corp., New York, USA.

V.NAGY, I. (1977) Application of Stochastic Methods in Hydraulics, *Periodica Polytechnica, Civil Engineering,* Vol. 21, No.3-4, Technical University, Budapest, Hungary.

V.NAGY, I. (1980) Design of Reservoir Capacity for Variable Water Demand, *Periodica Polytechnica, Civil Engineering,* Vol.24, No.3-4, Technical University, Budapest, Hungary.

V.NAGY, I. (1982) Prediction of the Transition Probabilities of Various Alert Stages during rising Flood. *Periodica Polytechnica, Civil Engineering,* Vol. 16, No. 4, pp.211-217, Technical University, Budapest, Hungary.

V.NAGY, I. (1990) Design of Reservoir Capacity for Variable Water Demand. *Periodica Polytechnica, Civil Engineering,* Vol. 24, No. 3-4, pp.261-266, Technical University, Budapest, Hungary.

V.NAGY, I., and BENEDICT SZ. (1990) *A Possible Risk Assesment Method in Flood Control Decisions,* IFAC/IFORS/imacs Symp. Publ., pp. 118-130, Tallin, Estonien.

VOGEL, R. and R. BOLOGNESE (1995) Storage-Reliability-resilience-Yield Relations for Over-Year Water Supply Systems, *Water Resources Research,* 31(3): 645-654.

WHO (1974) Guide to Hydrological Practices, Report No. 168, Geneva, Switzerland.

WILLIAMS, J.R. (1975) 'Sediment-yield prediction with the universal equation using runoff energy factor.' In, *Present and prospective technology for predicting sediment yields and sources,* U.S. Department of Agriculture, ARS-S-40.

WISMER, D.A. (ed.) (1971) *Optimization Methods for Large-Scale Systems,* McGraw-Hill, New York, USA.

WOOD, E.F. and I. RODRIGUEZ-ITURBE (1975) Bayesian Approach to Analyzing Uncertainty among Flood Frequency Models, *Water Resources Research,* 11(6), 839-843.

WURBS, R.A. (1995) *Water Management Models: A Guide to Software,* Prentice Hall, Englewood Cliffs, New Jersey, USA.

WURBS, R.A. (ed.) (1996) *Modeling and Analysis of Reservoir System Operations,* Prentice Hall, Upper Saddle River, New Jersey, USA.

WURBS, R.A., and C.E. BERGHMAN (1990) Evaluation of Factors Affecting Reservoir Yield Estimates, *Journal of Hydrology,* 112:219-235.

YAKOWITZ, S. (1982) Dynamic Programming Applications in Water Resources, *Water Resources Research,* 18(4), 673-696.

YAN, J. and B. FANWEI (1987) *Large Dams in China,* China Water Resources and Electrical Power Press, Beijing, China.

YEH, W.W.-G. and L. BECKER (1982) Multiobjective Analysis of Multireservoir Operations, *Water Resources Research,* 18(5): 1326-1336

YEVJEVICH, V. (1972a) *Probability and Statistics in Hydrology,* Water Resources Publication, Fort Collins, Colorado, USA.

YEVJEVICH, V. (1972b) *Stochastic Processes in Hydrology,* Water Resources Publication, Fort Collins, Colorado, USA.

YOUNG, G.K. (1967) Finding Reservoir Operating Rules, *Journal of Hydraulic Division, ASCE,* 93(HY6), 297-321.

YOUNG, G.K. (1968) Discussion of 'Mathematical Assessment of Synthetic Hydrology' by N.C. Matalas, *Water Resources Research,* 4(3).

ZSUFFA, I. (1979). Hydrological Design of Multiobjective Reservoirs and Reservoirs Systems, IIASA IIMGW Workshop, Warsawa, pp. 18-25.

ZSUFFA, I. (1982) *A Stochastic Approach to Reservoir Dimensioning: Summary,* Institute for Water Management, Budapest, Hungary.

ZSUFFA, I.J. (2000) Multi-criteria Decision Support for the Revitalisation of River Floodplains, Doctoral Dissertation, Wageningen University, The Netherlands.

ZSUFFA, I. and J. BOGARDI (1997) Non-stationary quasi-two-dimensional numerical model for hydrodynamic simulation of floodplain foksystems, *Journal of the Hungarian Hydrological Society,* 77(5): 369-279.

ZSUFFA, I. and A. GALAI (1987) *Reservoir Sizing by Transition Probabilities,* Water Resources Publications, Littleton, Colorado, USA.

ADDITIONAL SUGGESTED LITERATURE

Bedient, P.S. and W.C. Huber (1992) *Hydrology and Floodplain Analysis, 2nd ed.*, Addison-Wesley, Reading, MA.

Blom, C.W.P.M. and L.A.C.J. Voesenek (1996) Flooding: the Survival Strategies of Plants. *Trends in Ecology and Evolution*, 11:290-295.

Bogardi, I., A. Bardossy and L. Duckstein (1985) Effect of Parameter Uncertainty on Calculated Sediment Yield, *Advances in Water Resources*, 8: 96-101.

Bogardi, J.J. and A.A. Sutanto (1994) Discrete Decisions in Water Resources Planning: an Interactive Sequential Approach with Multiple Objectives. In: *J.J. Bogardi and H-P Nachtnebel, Multicriteria Decision Analysis in Water Resources Management*, UNESCO, Paris, France.

Bohac, C.E. (1989) *Environmental Issues and Mitigative Approaches*, ASCE/EPRI Guides, Chattanooga, Tennessee Valley Authority.

Bovee, K.D. (1986) *Development and Evaluation of Habitat Suitability Criteria for use in Instream Flow Incremental Methodology*, U.S. Fish and Wildlife Service Paper No. 21.

Brooks, K.N., P.F. Ffolliott, H.M. Gregersen and L.F. DeBano (1997) *Hydrology and the Management of Watersheds, 2nd ed.*, Iowa State University Press, Ames, Iowa, USA.

Central Board of Irrigation and Power (1978) *Forty-seventh Research Session Proceedings, Vol. 1 – Hydraulics*, Publication No. 132, Research Scheme Applied to River Valley Projects, New Delhi.

Chang, H.H. (1988) *Fluvial Processes in River Engineering*, John Wiley and Sons, Inc., New York.

Chen, W.F. (ed.) (1995) *The Civil Engineering Handbook*, CRC Press, Boca Raton, Florida, USA.

Chow, V.T., D.R. Maidment and L.W. Mays (1988) *Applied Hydrology*, McGraw Hill, New York.

CIGB-ICOLD (1988) *Sixteenth International Congress on Large Dams*, San Francisco USA.

Cooke, R., M. Mendel, and H. Vrijling (eds.) (1997) *Engineering Probabilistic Design and Maintenance for Flood Protection*, Kluwer Academic Publishers, Dordrecht, Holland.

CSCE (Canadian Society for Civil Engineering) (1979) *River Basin Management, Volumes I & II*, Proceedings, 4th National Hydrotechnical Conference.

Ditlevsen, O. and H.O. Madsen (1996) *Structural Reliability Methods*, John Wiley and Sons, Chichester, UK.

Dixon, J.A., L.M. Talbot, and G.J-M. LeMoigne (1989) *Dams and the Environment (Considerations in World Bank Projects), World Bank Technical Paper Number 110*, The World Bank, Washington, DC.

Dokulil. M.T. (1995) Long-term Nutrient Loading and Biological Response in a Flood-Water Reservoir, *Water Science and Technology*, pp. 15-25.

Dreyfus, S. and A. Law (1977) *The Art and Theory of Dynamic Programming*, Academic Press, Orlando, Florida.

Erickson, A.J. (1977) *Aids for Estimating Soil Erodibility – "K" Value Class and Tolerance*, U.S. Dept. of Agriculture, Soil Conservation Service, Salt Lake City, UT, USA.

Garg, S.K. (1985) *Hydrology and Water Resources Engineering*, Khanna Publ., New Delhi, India.

Gill, M.A. (1988) Planning the Useful Life of a Reservoir. *Water Power and Dam Construction*, May 1988, pp. 46-47.

Gnedenko, B.V. (1962) *The Theory of Probability*, Chelsea Publ. Co., New York, USA.

Goda L. (1990) Hydrological Sizing of Reservoirs for Flood Protection, *Proc. of Technical College for Water Management*, Vol. 2, No 3, pp. 9-18, Baja, Hungary.

Goldman, S.J., K. Jackson and T.A. Bursztynsky (1986) *Erosion and Sediment Control Handbook*, McGraw-Hill, N.Y., USA.

Gordon, N.D., *et al.* (1992) *Stream Hydrology: An Introduction for Ecologists*, Wiley, Chiechester, U.K.

Haan, C.T. (1977) *Statistical Methods in Hydrology*, Iowa State University Press, Ames, Iowa, USA.

Haith, D.A. (1980) A mathematical model for estimating pesticide losses in runoff, *Journal of Environmental Quality*, 9(3):428-433.

Haktanir, T. (1989) Storage-yield Relationships for Reservoirs by a Sequential-operation Algorithm and Comparison with Gould's Probability Matrix Method, *Journal of Hydrology*, 109:43-56.

HEC (1991) *Optimization of Multiple –Purpose Reservoir System Operations: A Review of Modeling and Analysis Approaches, Research Document RD-34,* U.S. Army Corps of Engineers, Hydrologic Engineering Center, Davis, Calif., USA.

Helsel, D.R. and Hirsch, R.M. (1992) Stochastic Methods in Water Resources, Elsevier Co., Amsterdam, Holland.

Hootsmans, M.J.M. and H.J. Lubberding (1993) *Ecosystem Analysis,* IHE, Delft, pp. 10-30.

Horvath, E. (2000) Duration and Future of the Large Dams of Water-reservoirs, *Journal of the Hungarian Hydrological Society,* Vol. 80, No. 3, pp. 192-205.

Hrelja, H. and Isailovic, D. (1990) A Generalized Model of the Required Storage Capacity, *Journal of Hydrology,* 117:153-167.

Jobin, W. (1999) *Dams and Disease (Ecological Design and Health Impacts of Large Dams, Canals and Irrigation Systems,* E&FN Spon/Routledge, New York, NY, USA.

Jovanovic, A.S. (1999) *Intelligent Systems in Engineering. Applications in the Area of Risk-based Life Management of Critical Components in Power and Process Plants,* Laboratic de Mecanique des Solides Ecole Polytechnique, 91128, Palaiseau, France.

Kao, J-J. (1999) *Non-point Source Pollution Control Strategy for an Off-stream Reservoir,* Publ. of the Chiao Tung Univ., Peking, China.

Karadi, G.M., V.Nagy I., (1993) Optimal Operation of the Gabcikovo-Nagymaros Hydropower System, *Proc. of the International Conference: "Hydropower Systems" Bratislava,* Slovakia, pp. 51-63.

Kjelds, J.T., T. van Kalken, and W.J. Syme (1996) Integrating MIKE 11 Flood Dynamics with ArcView GIS for Flood Defence and Flood Plain Management, *Proc. RIVERTEC H 96, IWRA,* Urbana, IL, USA, pp. 381-388.

Klemes, V. (1979) Storage Mass-curve Analysis in a System-analytic Perspective, *Water Resources Research,* 15(2): 28-45.

Kondo, S. and K. Furuta (eds.) (2000) Probabilistic Safety Assessment and Management, *Proc. of the 5th International Conference on PSAM,* Vol. 2/4, Univ. Acad. Press, Inc. Tokyo, Japán.

Kothyari, U.C., A.K. Tiwari and S. Ranvir (1994) Prediction of Sediment Yield, *Journal of Irrigation and Drainage Engineering,* 120: 1122-1131.

Kottegoda, N.T. (1980) *Stochastic Water Resources Technology,* Macmillan Press, London, UK.

Lloyd, E.H. (1993) The Stochastic Reservoir: Exact and Approximate Evaluations of the Storage Distribution, *Journal of Hydrology,* 151, pp. 65-107.

Lloyd, E. H. (1996) On the Integral Equation of the Continuous-state Stochastic Reservoir with Seasonally Varying Autocorrelated Inflows, *Journal of Hydrology,* 180, pp. 351-360.

Mays, L.W. and Y-K. Tung (1992) *Hydrosystems Engineering and Management,* McGraw-Hill, New York, NY, USA.

McCuen, R.H. (1989) *Hydrologic Analysis and Design,* Prentice Hall, Englewood Cliffs, New Jersey, USA.

Moore, J.W. (1989) *Balancing the Needs of Water Use,* Springer-Verlag, New York, NY, USA.

Morris, G.L. and J. Fan (1998) *Reservoir Sedimentation Handbook,* McGraw-Hill , New York, NY, USA.

Mosonyi, E. (1996) The Computation of the T-years Flood by Means of the Probable Maximum Flood (PMF) and Beta-Distribution, *Journal of the Hungarian Hydrological Society,* Vol. 76, No. 1, pp. 12-18.

Mosonyi, E. (1999) The Design Flood (in Hungarian), *Hydraulic Engineering,* No.2, pp. 201-215, Budapest, Hungary.

Moxon, J. (1969) *Volta: Man's Greatest Lake,* Frederick A. Praeger Publishers, New York, NY, USA.

Nagy, L. (2000) Raising the Level of Flood Safety, *Journal of the Hungarian Hydrological Society,* Vol. 80, No. 2, pp. 111-118.

Nalbantis, I. and D. Koutsoyiannis (1997) A Parametric Rule for Planning and Management of Multiple Reservoir Systems, *Water Resources Research,* 33(9): 2165-2177.

NRC (National Research Council) (2000) *Risk Analysis and Uncertainty in Flood Damage Reduction Studies,* National Academy Press, Washington, DC, USA.

Petts, G.E. and C. Amoros (eds.) (1996) *Fluvial Hydrosystems,* Chapman & Hall, London, UK.

Phien, H.N., A. Arbhabhirama and P. Chenyapanich (1980) Reservoir Sedimentation with Random Deposits, *Water Resources Bulletin,* 16(5): 818-823.

Pourahmadi, M. (2001) *Foundations of Time Series Analysis and Prediction Theory,* J. Wiley & Sons, New York, NY, USA.

Royal Institute of Technology (1986) *Hydrology for Planning and Construction*, Bulletin No TRITA-VBI-130, pp. 1-35, The Royal Institute of Technology, Stockholm.

Sabian, A.A. (1994) Reservoir Sizing by the Grid Method, *International Water Power and Dam Construction*, May 1994, pp. 40-43.
Sabian, A.A. (1994) Reservoir Sizing by the Grid Method, *International Water Power and Dam Construction*, May 1994, pp. 40-43.
Schnitter, N.J. (1994) *A History of Dams, the Useful Pyramids*, A.A. Balkema, Rotterdam, Holland.

Schwab, G.O., R.K. Frevert, T.W. Edminster and K.K. Barnes (1966) *Soil and Water Conservation Engineering, 2nd ed.*, New York: John Wiley & Sons

Singh, K.P. and A. Durgunoglu (1989) A New Method for Estimating Future Reservoir Storage Capacities. *Water Resources Bulletin*, 25(2): 263-274.

Stanners, D. and D. Bourdeau (1995) *Europe's Environment (The Dobris Assessment)*, European Environment Agency, Copenhangen, Denmark.

Stedinger, J.R., B.F. Sule and D. Pei (1983) Multiple Reservoir System Screening Models, *Water Resources Research*, 19(6): 1368-1393.

Tecle A. and L. Duckstein (1994) *Concepts of Multicriterion Decision Making*, UNESCO, Paris, France.

Thomson, K., J. Stedinger and D. Heath (1997) Evaluation and Presentation of Dam Failure and Flood Risks, *Journal of Water Resources Planning and Management*, July/August. pp. 28-36.

Todorovic, P. and E. Zelenhazic (1970) A stochastic model for flood analysis, *Water Resources Research*, Vol. 6, pp. 28-50.

Tundisi, J.G. (1998) Reservoir Management in South America, *International Journal of Water Resources Development*, pp. 25-35, June 1998.

USGS (1982) *Guidelines for Determining Flood Flow Frequency, U.S. Geological Survey, Bulletin #17B*, U.S. Dept. of the Interior, Office of Water Data Coordination, Reston, VA, USA.

V.Nagy, I. (1998) Problems in Assessing the Ecological Impacts of River Canalization, *Journal of the Hungarian Hydrological Society*, Vol. 71, No. 3, pp. 147-151.

Vogel, A. (2000) The influence of the dam type to the hazard of dam failures caused by human factors, *Proc. of Esrel 2000, Sars and Sra-Europe Annual Conference, Edinburgh/Scotland*, Offprint A.A. Balkema/Rotterdam/Brookfield, pp. 1369-1372.

Voortman, H.G. and J.K. Vrijling (2001) A Risk-based Optimisation Strategy for Large-scale Flood Defence Systems, *Proc. of the International Conference on Safety, Risk and Reliablity*, pp. 549-554.

Votruba, L. and V. Broza (1989) *Water Management in Reservoirs (Developments in Water Science, Vol. 33)*, Elsevier Science Ltd., Holland.

INDEX

A

Acceptable risk level
  selection of 110
Active storage 35-41
Autocorrelation function 134

B

Bayesian techniques 55-57

C

Classification of reservoir
  capacity-yield procedures 125-126
Coefficient of correlation 132-134
Critical period 126
Critical period techniques 126-129
Cross-correlation function 133-134

D

Dams
  classification of 4
  planning for 29-60
  purpose of 4, 11
  site selection for 31
Data generation
(*see* Stochastic data generation)
Design flood
  determination of 109
Disaggregation model 92, 152-157
Dynamic programming 103, 140

E

Economic concepts
  benefit-cost analysis 101
  cost-effectiveness analysis 101
  economic damage
   evaluation 112-113
  hydro-economic
   impact analysis 112-113
  marginal analysis 101
  in reservoir planning 100
  in water resources 100-105
Efficiency function
  of flood protection
   reservoirs 172-181
  of the mass-curve methods 129
  a probabilistic model 204-208
Emergency flood storage
  reservoir 182-183
  impacts of 183-184
  design and operation of 184-186
Equivalent flood storage capacity 43
Evaporation
  basic requirements 49-51
  estimation strategies 49
  loss adjustments 48-51

F

Flood
  definitions 17-19
  discharge 18
  duration 18
  elevation 18
  flow velocity during 18
  magnitude 17
  seasonal occurrence of 18
  volume 18

Flood analysis
  comparison of distributions 114
  definition 18, 42
  objective 18
Flood control 109
Flood flow
  computations 42
  design flood 43, 109
  estimation of attenuation
    by reservoirs 122-124
  estimation using limited data 122
  hydrologic analysis of 109-124
  modeling 113
  probability of occurrence 115-121
  return period 42
Flood-frequency curve 18

Flood hydrograph 9,19
Flood retention reservoirs
  determination of the characteristic
    hydrograph 169-170
  dimensioning 169-186
  spillway and sluice gate
    considerations 170-178
Flow allocations 15-16
Flow regulation
  annual 22-24
  daily 19-20
  the hydrology of 15-27
  multiannual 22, 24
  the need for 5-11
  optimizing 25
  reliability of 16-17
  scope and types 1, 5-8, 15, 19-24
  seasonal 22-24
  typical structures for 15
  using storage reservoirs 7, 13
  weekly 20-21
Freeboard
  definition 52
  purpose 52-53

G

Generated data
  88, 91-93, 126, 131-137
Gould's probability
  matrix method 126, 129

H

Hybrid models 95
Hydrological problems
  from water deficiency 16-17
  by water excess 17-19
Hydrologic design
  deterministic methods of 57-58
  guidelines 59-60
  key statistical parameters 54-55
  principles and concepts 61-84
  reliability-based techniques 61-72
  selecting a distribution 54-55
  stochastic methods of 57-59
  systems approach 85-107
  using inadequate data 90-93, 122
Hydrologic models
  Classification of 85-86
  selection, calibration,
    and use of 87-88
  time series 88-89
  types of synthetic flow 89
Hydrological sizing of reservoirs
  for flood protection 169-186
Hydrologic uncertainties
  categories 99
  decisions under 99

L

Linear programming 103, 140

M

Markov chain model
  82, 92, 187, 192-202
Mass curve method 126-129
Modeling
  analytical optimization models 89
  continuity constraints 75-77
  flood storage submodel 80-81
  general classes 89
  hydropower constraints 79
  irrigation constraints 78-79
  interbasin transfers 79
  probabilistic models 89

reservoir constraints 78
in reservoir design 75-81
simulation 90, 93
statistical techniques 90
techniques 89-90
Model verification 96
Monte Carlo analysis
basic requirements 151
techniques in optimal reservoir design 151
Monte Carlo model 151-157, 166
Moran model
application in reservoir storage design 187-208
a proposed model 190-204
Moran probability matrix procedures 129, 187-208
Multivariate stochastic models
annual flow models extension 93
model 152-156
disaggregation model 92
the multivariate Matalas model 93
Multireservoir systems
cost-efficient design 163-164
design of 125-138
isolated linked-up unit 149
outflows from 150
screening models 140-142
typical interconnection 148
Multisite multiseason flow generation strategy 151-157

N

Nonlinear programming 103

O

Optimal design
minimum release policy 81
maximum release policy 81
basis for optimum design 81
release policy 81-83
stochastic approach to 82-83
Optimization models 140
Optimization techniques
chance-constrained programming 107
dynamic programming 103, 140
linear programming 140
multiobjective 104
the net-benefit function 105
non-linear programming 103
the optimization problem 102
solution techniques 103-105
stochastic programming 106
Optimal reservoir storage capacity
determination of the 139-168
Optimal system design 139, 165-168

P

Probability distribution functions
Beta distribution 116-118
estimation of maximum floods 115-121
Gaussian distribution 116
in the Moran model 187-189
types 114
Probability matrix methods
basic principles 130
general types 129-131
Gould procedure 129
Moran-related techniques 129
Probable maximum flood (PMF) 110, 113
Probable maximum precipitation (PMP) 113

R

Reliability parameters
performance reliability 95
risk of failure 95, 97
supply reliability 70-71
temporal reliability 68-69
types 68-71, 97-99
Reliability programming 140-141
Reservoirs
characterization of 30
design of 1, 14, 30
design of multipurpose 32-33

hydrologic design elements 29-60
impacts of 11-13
independently acting 33-34
jointly acting 34
multiple 33-35
the need for 1
network of compensating 35
operational characteristics 29-60
parallel 33
planning for 29-60
purpose 11, 26, 35
relative capacity of 137-138
the role of hydrology 26, 31
siltation of 44-48
site selection for 31
storage capacity of 15
storage zones 36
tandem 33
topographical characteristics 72-75
total inflow 147
trap efficiency of 47-48
the water balance of 8
Reservoir capacity design 141-142
Reservoir capacity-yield estimation
based on stochastic data 131-137
classification 125-126
critical period techniques 126-129
by mass-curve procedure 127-129
minimum flow approach 127
probability matrix techniques 129
procedures 125-138
range 127
Rippl's method 127, 190
sequent-peak algorithm 127
Reservoir efficiency function
basis for storage determination 65-67
concept of 62-65
defined 61-71
Reservoir network analyses 142-146
Reservoir performance reliability
analysis 95-100, 130-131
annual reliability 98
reliability characteristics 97-99
reliability measures 68-71, 97-99
time reliability 98
volumetric reliability 98
Reservoir storage capacity
allocation of 37-39
components 35
determination of 35-53
multipurpose 35-36
size selection factors 35-53
Return period 111
Risk
performance reliability 95
reliability 95
in reservoir design 95-100
risk of failure 95, 111
risk of overtopping 111-112

S

Seepage losses 51-52
Sequent-peak algorithm 127
Space rule 142
Stochastic data generation
(*see* Synthetic flow data)
Stochastic reservoir models 140
Stochastic simulation
Annual Markov model 92
data generation scheme 93-95, 131-137, 151-157
defined 91-93
model validation 96-100
multivariate stochastic models 92-93
Thomas-Fiering Seasonal Model 91
Storage
active 35-41
adjustments for evaporation 48-51
allocation 36-53
capacity design methods 125-138
conservation 33-41
controlled vs. uncontrolled 43-44
dead 35-39
flood control 35-39, 41-44
for flow regulation 9, 13, 40
for hydroelectric power 41
sediment reserve 44-48
for water supply 40
Storage capacity allocation to reservoir sites 157-163

Synthetic flow data
  flow generation strategies 91-93
  need for 91-92
  stochastic generation of 91-93

T

Target flow allocation 16
Time series models
  characteristics of 88
  persistence 88
  stationarity 88
  uses 89
Transboundary river management 10
Transition probability 82, 187-208
Transition probability matrix 187-209

U

Uncertainty
  assessment of 99
  optimization under 106-107
  in reservoir design 95-100

W

Water balance equation 8, 190
Water resources
  management 2
  planning and design 3
  systems analysis 3

Y

Yield model 140

MIX
Papier aus verantwortungsvollen Quellen
Paper from responsible sources
FSC® C105338
FSC
www.fsc.org

If you have any concerns about our products,
you can contact us on
ProductSafety@springernature.com

In case Publisher is established outside the EU,
the EU authorized representative is:
Springer Nature Customer Service Center GmbH
Europaplatz 3, 69115 Heidelberg, Germany

Printed by Libri Plureos GmbH
in Hamburg, Germany